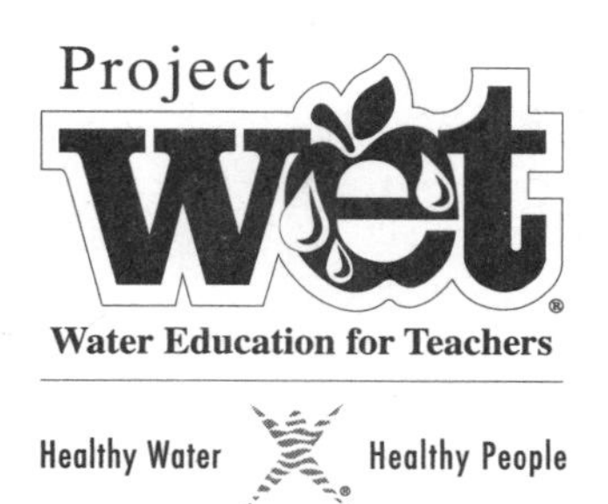


# Healthy Water, Healthy People Field Monitoring Guide

2114 North Lincoln
Loveland, Colorado 80538

Project WET International Foundation
1001 W. Oak Street, Suite 210
Bozeman, MT, USA 59715
www.projectwet.org

Cover photo credits:
Clouds © Don Farrall
Sawtooth Mountains, ID © Dick Dietrich
Ocean Wave © Digital Vision
Proxy Falls, OR © Dick Dietrich
Rock Dwelling © John P. George

ISBN: 1-888631-13-9

First Printing 2003
Second Printing 2006



The Project WET International Foundation is an award-winning nonprofit science, natural resources, and heritage education program and publisher in Bozeman, Montana, U.S.A.

Project WET, a nonprofit water resources education program, is located in Bozeman, Montana U.S.A. The mission of Project WET is to reach children, parents, educators, and communities of the world with unbiased water education. Project WET's goal is to provide scientifically accurate and educationally sound water resources education materials, training courses, and networking services to water and education agencies/organizations for use in designing, developing, and implementing their own localized Project WET programs based on the Project WET model. Project WET partners with agencies/organizations that share its mission, goal, and the following core beliefs:

- Water is important to all water users (e.g., business and industry, earth systems, energy, fish and wildlife, navigation/transportation, recreation, rural and agricultural, and urban/municipal)
- Wise water management is crucial for providing tomorrow's children social and economic stability in a healthy environment
- Awareness of, and respect for, water resources can encourage a personal lifelong commitment of responsibility and positive community participation

Established in 1989, Project WET works with visionary sponsors, educators, resource professionals, business leaders, policy makers, and citizens in the creation, development, and implementation of their projects. Project WET responds to the needs of many diverse groups and relies on public and private partnerships to accomplish its work.

Project WET reaches millions of people each year through their international delivery network. This extensive grassroots support is a hallmark of Project WET.

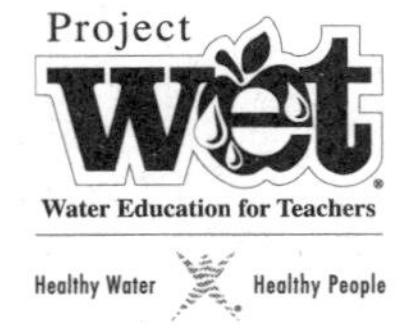


**Word Usage, Grammar, and Writing Style**

The writing style within this guide follows the Chicago Manual of Style, 14th edition; spelling is based on the Random House Unabridged Dictionary, 2nd edition. The term ground water is presented as two words based on the recommendation of the United States Geological Survey (USGS), the primary water management data agency for the country.

Printed on recycled paper.

# Dedication

Dear Educator:

The primary support for Healthy Water, Healthy People came from the Kathryn Hach Trust. Clifford Hach was dedicated to the science of quantitative analysis and developing tests and methods for the precise measurement of known constituents in water. Hach tests are complete, simplified, accurate and employed by water management professionals around the world.

There can be no life without water. The water management industry deeply probes analysis and fundamental characterization of water's properties.

Healthy Water, Healthy People will introduce fundamentals of scientific exploration, and work with the unique character of water chemistry and water behavior. The "good wishes" of this book are dedicated to the competent commitment of water management professionals and the dedication of Hach Company chemists, scientists and associates.

Kathryn Hach-Darrow
Co-Founder/Former Owner of Hach Company
Chairman
Hach Scientific Foundation

The mission of the Hach Scientific Foundation is to make evident the interdependence between science education and the public.

# Preface

*John Etgen, Director*
*Healthy Water, Healthy People Program*

Water quality has been a priority topic of Project WET since the program's inception in 1985. Between 1991 and 1995, Project WET conducted a national curriculum-writing project with peer-nominated educators and water resources professionals representing all 50 states and several territories. These outstanding educators were asked to submit a list of three of the most important water topics that Project WET should pursue to supplement the core water resources education guide. This national survey resulted in the identification of five priority topics:

- Watersheds and Rivers
- Wetlands
- Water History/Environmental History
- Ground Water
- **Water Quality, Environmental, and Public Health**

Project WET is committed to raising funds from public and private sources to publish education materials and delivery programs on each topic. Healthy Water, Healthy People is a direct result of Project WET's national and international need for a water quality education component.

### Why Healthy Water, Healthy People?

The purpose of this publication is to raise educators' awareness and understanding of water quality topics and issues by demonstrating the relationship of water quality to personal, public, and environmental health. This publication–especially when used in combination with the other Healthy Water, Healthy People materials —gives teachers, students, nonformal educators, water managers, treatment plant operators, and citizens an opportunity to explore water quality topics in an interactive, easy-to-use, hands-on format.

### What is Healthy Water for Healthy People?

Healthy water is simply water that supports and sustains life. All living things use water. It allows our food to grow, the trees to transpire, and our bodies to perform. The quality of the water is affected by a variety of factors, both natural and human-related. How healthy this water must be depends on how it is used.

The United States Environmental Protection Agency employs a classification of "designated uses" to determine what level of health water must attain. Water quality standards, or allowable levels of contaminants, are assigned to each of these uses. The most stringent standards apply to drinking water used in public water supplies. Following that, in order, are fish and wildlife, recreation, agriculture and industry, navigation, and other uses (e.g., hydroelectric, marinas, ground water recharge,etc.).

Meeting these water quality standards requires testing and measurement of the contaminant levels, which in turn requires a broader understanding of the sources, interactions, and remedies for these contaminants. Healthy Water, Healthy People publications and materials illustrate and promote the broad concepts of water quality testing and monitoring for educational purposes, while encouraging those who wish to become more involved to contact their local water quality monitoring program leader or state water quality specialist.

### Our Goal

The goal of Healthy Water, Healthy People is to make the complex concepts of water quality relevant and meaningful for you and those you teach. Please let us know how we are doing.

# Table of Contents

# Using the *Healthy Water, Healthy People Field Monitoring Guide*

**An Example of How the Publications and Testing Kits Work Together:** The Healthy Water, Healthy People Field Monitoring Guide highlights eleven common water quality parameters and is designed to be used as a technical water quality reference that supports the Healthy Water, Healthy People Water Quality Educators Guide and the Healthy Water, Healthy People Testing Kits. This manual could also serve as a general water quality text for environmental science, water quality, and hydrology courses.

### Challenge I: Classroom Teachers

An educator wants to teach her students about nonpoint source pollution, nutrient loading, and the resulting eutrophication.

### The Solution:

Using Healthy Water, Healthy People Materials:

1. The instructor begins the unit by conducting "Mapping It Out; Water Quality Concept Mapping" activity from the *Healthy Water, Healthy People Water Quality Educators Guide.* Through this activity, the students build a concept map using their prior knowledge of nonpoint source pollution, nutrient loading, and eutrophication.

2. She then conducts interactive activity, "There Is No Point To This Pollution!" from the *Healthy Water, Healthy People Water Quality Educators Guide.* This data analysis and mapping activity illustrates how nonpoint source pollutants and nutrients can enter waterways and accumulate, causing eutrophication.

3. A student then asks how phosphates are related to phosphorus, so she consults the "Phosphate" chapter in the *Healthy Water, Healthy People Field Monitoring Guide* for the answer (or asks the student to search for it).

4. She then instructs her students to conduct a phosphate study of their local lake using a *Healthy Water, Healthy People Testing Kit* to test the phosphate levels.

5. She then conducts a demonstration of how nutrients move through the soil using the "Testing Kit Activities" section of the *Healthy Water, Healthy People Field Monitoring Guide.*

6. She then assigns her students to conduct online research originating with the Healthy Water, Healthy People Web site–www.healthywater.org.

7. She concludes by revisiting the original concept map from the start of the unit. The students add to the map while the instructor assesses the knowledge gains between the beginning and the end of the unit.

# Using the *Healthy Water, Healthy People Field Monitoring Guide*

### Challenge II: Watershed Monitoring Project Leaders

The leaders of a water quality monitoring group wants their group to understand the WHY and HOW behind water quality monitoring of their local.

### The Solution:

Using Healthy Water, Healthy People Materials:

1. The instructors begin the unit by conducting "A Snapshot in Time" activity from the *Healthy Water, Healthy People Water Quality Educators Guide.* This activity gives the students a foundation in understanding watersheds, data collection, and analysis.

2. They then conduct interactive activity, "Hitting the Mark" from the *Healthy Water, Healthy People Water Quality Educators Guide.* This accuracy and precision activity gives the students an introduction into the importance of these concepts in relation to data collection.

3. A third activity, "Water Quality Monitoring; From Design to Data" is conducted by the instructors and the students are introduced to the fundamentals of study design in relation to a water quality monitoring project. The students also learn to analyze and compare real data.

4. A student then asks what can affect the dissolved oxygen content of a river, so the instructors consult the "Dissolved Oxygen" chapter in the *Healthy Water, Healthy People Field Monitoring Guide* for the answer (or asks the student to search for it).

5. The instructors then ask the students to design and conduct a water monitoring study of their local river using the *Healthy Water, Healthy People Rivers, Streams, Ponds, and Lakes Testing Kit.*

6. They then conduct a demonstration of how dissolved oxygen is consumed by using the "Testing Kit Activities" section of the *Healthy Water, Healthy People Field Monitoring Guide.*

7. The instructors then ask the students to conduct online research originating with the Healthy Water, Healthy People Web site–www.healthywater.org.

# Using the *Healthy Water, Healthy People Field Monitoring Guide*

## Parameter

**Primary Importance**
The most critical water quality application of the parameter.

**Technical Overview And History**
Technical discussion explaining the background and history of the parameter; includes chemical formulas and other highlights.

**Sources**
Explains where the parameter or contaminant originates.

**Relevance**
The application of the parameter to humans, animals, and plants.

**Impacts**
How the parameter affects humans and the environment.

**Possible Remedies**
Sample management strategies and everyday actions that can help remediate the impacts from the parameter.

**Case Study**
Explains a real-life situation where issues related to the parameter were managed or remediated.

**Test Kit Activities**
Hands on activities or instructor demonstrations that use a water quality testing kit to illustrate concepts related to the parameter.

**Parameter Summary**
A summary outline of the primary elements of the parameter; includes Sources, Relevance, Impacts, and Remedies.

**Indicates Test Kit Activity**

**Visit www.healthywater.org For More Information**

**Indicates Student Copy Page**

**References**
Bibliography of the references used in the chapter.

**Appendixes**

**A. Interpreting Your Water Quality Data:** A quick reference tool which gives possible causes, effects, and expected ranges for each parameter.

**B. Cross Reference Between Healthy Water, Healthy People Publications and Testing Kits:** A detailed reference chart illustrating the connections between the *Healthy Water, Healthy People Educators Guide, Healthy Water, Healthy People Field Monitoring Guide,* and the Healthy Water, Healthy People Testing Kits.

**C. A Brief History of Chemistry Up to the Discovery of the Water Molecule:** An essay on the history of chemistry by Bruce J. Hach; Managing Director of the Hach Scientific Foundation.

**D. Periodic Table**

**E. Metric Conversions**

**F. Glossary:** Definitions of the technical terms from the manual.

**G. Index:** Page numbers associated with the topics/terms from the manual.

**H. Milestones in Water Quality Management**

# Healthy Water, Healthy People Program Overview

**Water Quality Education Program**
The Hach Scientific Foundation and Project WET (Water Education for Teachers), drawing on over fifty years of success in water quality test kit manufacturing, service, and water education, have partnered to create Healthy Water, Healthy People—a new and innovative water quality education program. Healthy Water, Healthy People is associated with Project WET, and originated as a result of a growing need for information and education on water quality, not only in the United States, but also around the world. The Healthy Water, Healthy People program encourages deep investigation of water quality topics and issues through development of user-friendly materials that are appropriate for all levels of users—from beginner to advanced.

Healthy Water, Healthy People recognizes that clean water is important for all people, prosperous economies, and natural systems. Water education must play an important role in providing opportunities for all citizens to learn about water quality in ways that are relevant and meaningful. Understanding the relationship of healthy water to healthy people will be critical as we collectively work to develop solutions for addressing continued water quality challenges and opportunities.

**Mission Statement**
The mission of Healthy Water, Healthy People is to reach children, young adults, educators, parents and communities with water quality education.

**Goal**
The goal of the Healthy Water, Healthy People Program is to facilitate and promote the awareness, appreciation, knowledge, stewardship, and understanding of water quality topics and issues, and to make evident the interdependence between science education and the public.

**Audience**
Healthy Water, Healthy People is for anyone interested in learning and teaching about contemporary water quality education topics:

- Upper Elementary through Secondary School Teachers
- Science Methods, Science Education and Environmental Science Professors
- River and Lake Monitoring Program Leaders, Drinking Water and Waste Water Facility Operators and Educators, Land and Water Managers, Conservation District and Extension Agents, Urban Program Members, Health Care Educators and Providers, Scien-tists, and Policy Makers
- Citizens–anyone interested in water quality

**The Program**
Healthy Water, Healthy People is an innovative, contemporary, and comprehensive water quality education program designed for anyone interested in learning about or teaching water quality. Healthy Water, Healthy People includes publications, testing kits, training and professional development, networking and support services, and an awareness campaign.

**Publications**
***Healthy Water, Healthy People Water Quality Educators Guide***
This 248-page activity guide is for educators of students in grades six through university level. The purpose of this guide is to raise the awareness and understanding of water quality topics and issues and their relationship to personal, public, and environmental health. Healthy Water, Healthy People will help educators address science standards through interactive activities that interpret water quality concepts and promote diverse learning styles, with foundations in the scientific method. This guide contains twenty-five original activities that link priority water quality topics to real-life experiences of educators and students.

***Healthy Water, Healthy People Field Monitoring Guide***
This technical reference manual is an excellent companion text that supports all of the *Healthy Water, Healthy People* publications and materials. The purpose of this manual is to serve

as a technical reference for the Healthy Water, Healthy People Water Quality Educators Guide and the Healthy Water, Healthy People Testing Kits, yielding in-depth information about eleven water quality parameters. The manual answers questions about water quality testing using technical overviews, data interpretation guidelines, case studies, chemical formulas, testing kit activities, laboratory demonstrations, and much more.

***Healthy Water, Healthy People KIDS (Kids in Discovery Series) Activity Booklet***

This colorful, fully illustrated 16-page activity booklet is for fourth through seventh grade students. Through informative text, activities, investigations, and experiments, the Healthy Water, Healthy People KIDS booklet is designed to illustrate water quality topics and issues and make them intriguing, relevant, and fun for kids.

**Testing Kits**

Developed in cooperation with the Hach Company, a worldwide leader in water quality testing equipment, these water quality testing kits include all materials and equipment needed for field and classroom analysis of water samples. The testing kits are cross-referenced with the above publications and allow students to conduct in-depth investigations and analysis of water quality parameters and issues. Healthy Water, Healthy People Testing Kits are available for a variety of parameters, grade levels, skills, and prices. Each Healthy Water, Healthy People testing kit package includes the *Healthy Water, Healthy People Water Quality Educators Guide*, a Testing Kit, and the *Healthy Water, Healthy People Testing Kit Manual.* To order a Healthy Water, Healthy People Testing Kit Package, contact the Healthy Water, Healthy People headquarters toll free at 1-866-337-5486, order online at www.healthywater.org, or contact us by email at info@projectwet.org. For Testing Kit technical support, contact Healthy Water, Healthy People.

**Healthy Water, Healthy People FirstStep™**

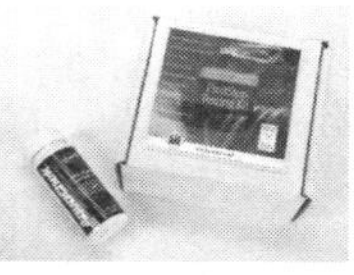

**Healthy Water, Healthy People Watershed Testing Kit**

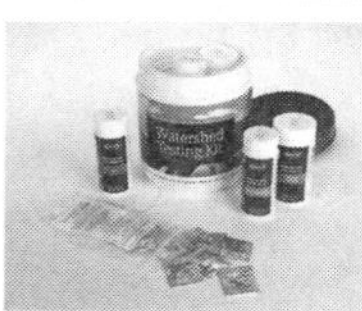

**Healthy Water, Healthy People Drinking Water Testing Kit–Urban, Rural, School**

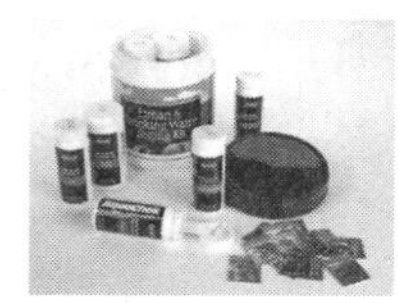

**Healthy Water, Healthy People Classroom Drinking Water Testing Kit**

**Healthy Water, Healthy People Rivers, Streams, Ponds, and Lakes Testing Kit**

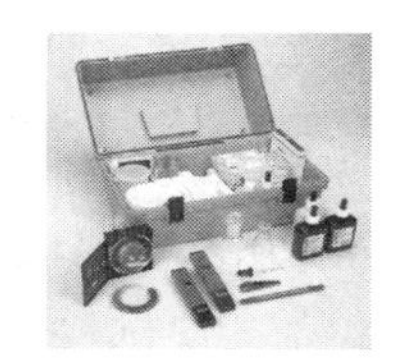

**Healthy Water, Healthy People Advanced Water Quality Testing Kit**

**Healthy Water, Healthy People Macroinvertebrate Investigation Kit–MacroPac™**

**Training Seminars, Workshops, and Institutes**

Healthy Water, Healthy People provides customized water quality training opportunities for educators, corporations, communities, government agencies, educational conferences, and non-governmental institutions. People and organizations interested in working with the Healthy Water, Healthy People program to sponsor training seminars, workshops, and professional development institutes for staff and constituencies should contact Healthy Water, Healthy People to discuss dates and locations. A national and international network of water quality education trainers will be established. The program will help people and organizations design, develop, and implement their own water quality education programs. To schedule training, contact the Healthy Water, Healthy People staff toll-free at 1-866-337-5486.

### Networking and Support Services

Healthy Water, Healthy People is committed to assisting people in their water quality education efforts. Staff members have extensive experience in water quality education programming and can help education providers answer questions.

•Education support–through Healthy Water, Healthy People staff 1-866-337-5486

•Network Communications Support–Healthy Water, Healthy People Newsgroup, subscribe at www.healthywater.org. This rapidly developing international network includes scientists, water educators, classroom teachers, water monitoring program leaders, healthcare professionals, and others interested in contemporary water quality education topics and issues.

### WE CARE!® Campaign

If you care about healthy water and healthy people and want to demon-strate this commitment while contributing to the public's understanding of this critically important topic, the WE CARE!® Campaign is for you. The WE CARE!® Campaign is a national and international water quality awareness initiative designed for public and private organizations and corporations to highlight their efforts to manage, protect, and restore water resources. There are amazing efforts taking place all around the world to provide people with clean and abundant water supplies and to sustain natural systems. Healthy Water, Healthy People staff can help these organizations demonstrate their support for healthy water in creative and customized ways. Contact the Director for more information.

### Sponsors

The Healthy Water, Healthy People program is sponsored by Project WET and was initially funded by a generous donation from the Hach Scientific Foundation.

The mission of the Hach Scientific Foundation is to foster and support science and science education, and to make evident the interdependence between science education and the public.

Financial support for additional Healthy Water, Healthy People projects has been provided by the Nestle´ Waters of North America, the Hach Company, and Project WET USA Network Sponsors.

### Sponsorship and Contact Information

Individuals or organizations interested in sponsoring Healthy Water, Healthy People should contact:

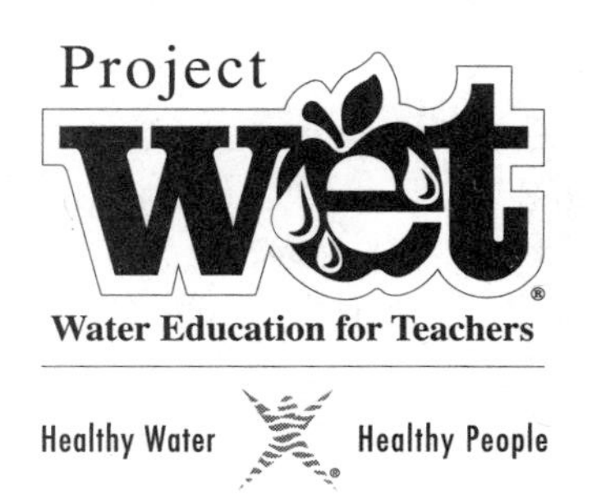


Director
Healthy Water, Healthy People Program

**Project WET**
1001 West Oak Street
Suite 210
Bozeman, Montana USA
59715

866-337-5486
Fax 406-522-0394

www.projectwet.org

# Acknowledgments

**Sponsors**
The *Healthy Water, Healthy People Field Monitoring Guide* was published through a partnership between the Hach Scientific Foundation and International Project WET. The mission of the Hach Scientific Foundation is to foster and support science and science education, and to make evident the interdependence between science education and the public.

**Publication Team and Contributors**
**Publisher**
Dennis L. Nelson
**Project Manager**
John E. Etgen
**Primary Authors**
John E. Etgen
Keri Garver
**Contributing Authors and Researchers**
Dennis L. Nelson
Bruce J. Hach
Ann DeSimone
Cassie Murray
Linda Hveem
Erynne Dues Joyner
Tawni Thayer
Savannah Barnes
Sandra DeYonge
**Production Coordinators**
John E. Etgen
Ivy Davis
**Editors**
Anne Taylor
John Richardson
Savannah Barnes
**Designer**
Ivy Davis, *Studio I.D.*
**Logo Design**
Ivy Davis, *Studio I.D.*
**Financial Management**
Stephanie Ouren

**Project WET International Foundation Staff**
Dennis Nelson
*President and CEO*
John Etgen
*Vice President for Development and Partnerships/Executive Director Healthy Water, Healthy People Program*
Gary Cook
*Vice President of Education/ Executive Director Project WET U.S.A.*
Meg Long
*Chief Financial Officer*
Linda Hveem
*Office Manager*
Rab Cummings
*National Network Coordinator, Project WET U.S.A.*
Scott Frazier
*Executive Director, Project WET's Native Waters*
Justin Howe
*Coordinator, Latin America & Caribbean Region/Director Project WET's Children's Education Fund*
Stephanie Kaleva
*Marketing and Communications Manager*
Elisabeth Howe
*Project Manager*
Sally Unser
*Internal Communications Coordinator Project WET U.S.A.*
Susan Denson
*Product Development Coordinator*
Verna Schaff
*Accountant*
Lindsey Lemon
*Customer Service and Front-Desk Representative*
Nick Johnson
*I.T. Coordinator*
Chelsea Goddard
*Work Study Accounting Assistant*
Leo Schlenker
*Warehouse/Shipping Assistant*
Meagan Hansen
*Work Study Office Support*
Crescentia Cummins
*Project WET's Native Waters Intern*

**Reviewers**
Hach Scientific Foundation Staff
Tom Aspelund, Hach Company
Denton Slovacek, Hach Company
Barbara Martin, Hach Company
Eric Umbreit, Hach Company
Janet Vail, Grand Valley State U.

# Alkalinity

## Primary Importance

Alkalinity is a property of water that helps prevent drastic changes in pH, and thus protects humans, wildlife, and aquatic life.

## Technical Overview And History

Alkalinity is a measure of water's capacity to resist a decrease in pH. It is also referred to as the acid neutralizing capacity, and sometimes the buffering capacity. Although a true buffer is able to neutralize both acids and bases, alkalinity is a measure of the concentration of bases that neutralize acids.

Alkalinity in natural waters is primarily a function of the carbonate system. Carbonates come from limestone and other rocks containing calcium carbonate that dissolve on contact with water. They release calcium ions ($Ca^{2+}$) and carbonate ions ($CO_3^{2-}$), bicarbonate ions ($HCO_3^-$), or carbonic acid ($H_2CO_3$), depending on the water's pH. The negative carbonate and bicarbonate ions then combine with the positive hydrogen ions ($H^+$) from solution, thereby reducing the acidity and increasing the pH (see pH). Different carbonate species dominate at different pH levels as listed below.

**Sources**

Natural

- Carbonate rocks

Human

- Wastewater treatment plant effluent

**Relevance**

Human

- Buffers acid inputs
- Removes heavy metals from solution

Animals

- Buffers acid inputs
- Removes heavy metals from solution

Plants

- Buffers acid inputs
- Removes heavy metals from solution

**Impacts**

Human health

- Low: lose buffering capacity
- High: corrosive to skin and pipes

Environmental health

- Low: lose buffering capacity
- High: harmful to aquatic life
- Should be at least 20 mg/l to protect aquatic life

**Possible Remedies**

- Remediate acid mine drainage and acid precipitation

**pH > 10.33**

carbonate ($CO_3^{2-}$) is the dominant species

**pH 6.4 to 10.33**

bicarbonate ($HCO_3^-$) is the dominant species

**pH < 6.4**

carbonic acid ($H_2CO_3$) is the dominant species

To determine the total alkalinity of water, a titration is performed in which a known quantity of acid is added to water until it reaches a pH of about 4 (Drever, 1997). The greater the volume of acid required to reduce the pH to 4, the higher the alkalinity in the water sample. The lower the volume of acid required to reduce the pH to 4, the lower the alkalinity.

If alkalinity is present, the pH will drop slowly to 6.4 as calcium carbonate combines with the acid titrant's $H^+$ ions to form bicarbonate. As the titration continues, pH drops rapidly, as most of the available acid neutralizing capacity is depleted, until a pH of about 5.2 is reached. At this point, available bicarbonate combines with the acid's $H^+$ ions to form carbonic acid. Finally, at a pH of about 4, all of the alkalinity is 'used up'—carbonate and bicarbonate ions are all converted to carbonic acid—and the water can no longer neutralize acid. If no alkalinity is present in the original water sample, the pH will drop quickly when acid is added.

## Sources

Alkalinity is a measure of all of the available bases in water, many of which come from carbonate minerals in sedimentary rocks. Limestone rock yields calcium carbonate, and dolostone yields dolomite or calcium and magnesium carbonate. Rainwater slowly weathers (dissolves) or leaches the minerals from these rocks and releases carbonate and bicarbonate compounds into the water. In areas where granite or other igneous rocks dominate the geology, waters have little natural alkalinity.

Green plants impact the carbonate system, and therefore water alkalinity, during photosynthesis by consuming $CO_2$ that would otherwise form carbonic acid ($H_2O + CO_2 = H_2CO_3$) when dissolved in water.

Human activity also influences the alkalinity of waters. Treated wastewater effluent contains cleaning agents made from carbonate and bicarbonate used in our hospitals, homes and businesses. Wastewater also contains residues from food substances, particularly vegetables and beans, that contribute to alkalinity.

*Helicopter drops lime in dying Lake Ovre to neutralize acid rain effects. Bergsjon, Sweden. Mark Edwards/Still Pictures/Peter Arnold, Inc.*

## Relevance

Alkalinity in natural levels is beneficial to all organisms that depend on water. Because alkalinity resists a change in pH it helps prevent acidic water (pH < 5) that is harmful to humans, wildlife, and aquatic organisms (see pH). Some acidic water also mobilizes toxic heavy metals, making them available to the environment. Alkaline compounds not only neutralize acidity, but also react with heavy metals, such as lead, arsenic, and cadmium, to remove them from the water.

## Impacts

Similar in effect to the antacid tablet that relieves acid indigestion, alkalinity counters the effects of industrial effluent and acid rain. With insufficient alkalinity, even small amounts of acid added to water will lower pH to an extent harmful to trees and wildlife, and fatal for aquatic organisms (see pH). Water alkalinity of at least 20 mg/L $CaCO_3$ will protect and sustain aquatic life (KRAMP, 1997).

Yet, too much alkalinity can be harmful to humans and aquatic organisms. Highly basic waters can be as corrosive as highly acidic waters. An absolute concentration of alkalinity can have different corrosivity effects, depending on the amount of acid present to counter it. Therefore, instead of establishing a quantitative maximum alkalinity level, the total alkalinity should not allow the pH to exceed 9.

## Possible Remedies

Generally, human inputs of alkalinity are not causing serious environmental problems. Of greater concern is the lack of alkalinity in areas that receive significant acid precipitation. Areas such as the northeastern United States do not have adequate carbonate rocks to mitigate the impacts of acid deposition (see pH).

To remedy this problem, sulfur and nitrogen emissions from automobiles and fossil-fuel-burning power plants could be reduced. Fluidized bed combustion in the fossil-fuel burning system adds alkaline materials (calcium rich limestone) to the fuel prior to the burning (Hem, 1970). This added limestone captures sulfur and nitrogen before it is emitted into the atmosphere. In some cases, blocks of carbonate rocks are added to lakes in northeastern United States to reduce the impacts of acid rain. Carbonate is also being added during mining processes to offset the impact of acid mine drainage.

www.

### Case Study

Pyrite ($FeS_2$) naturally occurs in coal. Left undisturbed, pyrite slowly undergoes chemical reactions via the natural weathering process to form small amounts of sulfuric acid. Natural alkalinity in the system can buffer the effects of this slow introduction of acid.

However, in coal mining operations in Pennsylvania for example, pyrite has been discarded in large quantities of rubble or mine spoil resulting in more acid drainage than can be buffered by natural alkalinity. In fact, most stream pollution in Pennsylvania comes from acid mine drainage (Bisko, 1998). Fortunately, in 1968, Pennsylvania began to limit the amount of acid effluent from mining operations (PA DEP, n.d.).

To meet new effluent regulations, mining operations began adding carbonate ($CaCO_3$) to drainage waters. However, this method of "active treatment" was expensive and required daily maintenance. Thus began the quest for more practical acid mine drainage remediation technology.

By the late 1970s, scientists were testing potential passive treatments that would create controlled conditions for natural biological and chemical remediation reactions. In general, these passive treatment systems add alkalinity to the acid drainage, and then allow the heavy metals to precipitate out of solution in a contained wetland before being discharged into the natural waters.

In 1999, several passive treatment systems were deployed in the South Fork of the Tangascootack Creek in Pennsylvania. Prior to implementation, the biological and chemical characteristics of the watershed were evaluated. Field studies of the Tangascootack and nearby tributar-

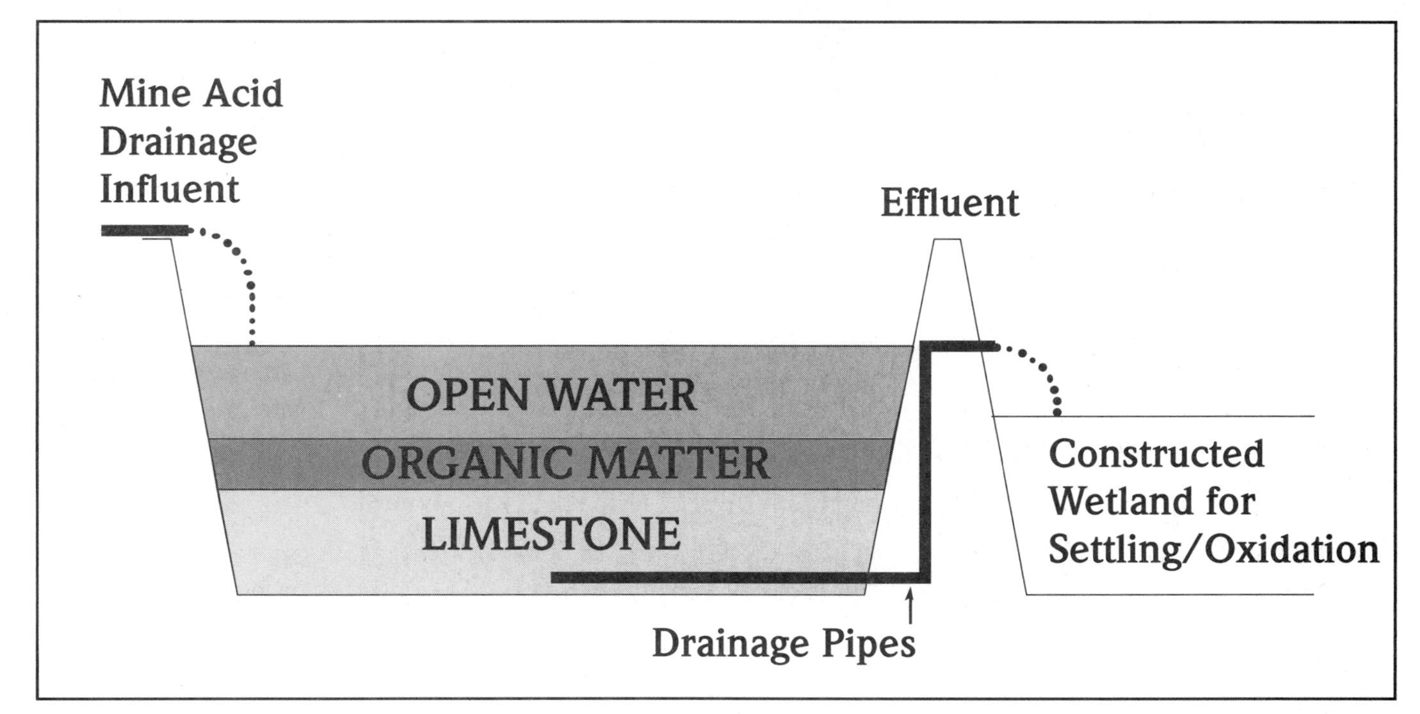


*Vertical Flow System also known as Successive Alkalinity Producing Systems, for Renovation of Acid Mine Drainage*

ies documented struggling or nonexistent aquatic life in waters with a pH as low as 4.3, and high concentrations of iron, manganese, aluminum, and sulfate (Bisko, 1998).

Specific data from each location helped guide selection of the appropriate passive treatment system. For example, waters with dissolved oxygen, ferric iron ($Fe^{3+}$), and aluminum concentrations less than 1.0 mg/l can be successfully treated with Anoxic Limestone Drain systems. Waters with concentrations of these constituents greater than 1.0 mg/l must be treated with a more aggressive technology, such as Vertical Flow Systems, also known as Successive Alkalinity Producing Systems (SAPS).

Anoxic Limestone Drains are underground limestone beds. When acid mine drainage reacts with the limestone, it increases the pH and total alkalinity of the system. Then, the alkaline water is discharged into a constructed wetland where it reacts with the atmosphere to oxidize and precipitate the heavy metals. The pH of waters from the South Fork of the Tangascootack Creek increased from pH 4.3 to pH 7.1. Alkalinity increased from 8 mg/l to196 mg/l with the use of an Anoxic Limestone Drain (Bisko, 2002).

Vertical Flow Systems include a network of drainage pipes covered by a layer of limestone gravel. Atop the limestone is a layer of organic material such as spent mushroom compost. Acid mine water is diverted to this treatment cell, or a series of treatment cells, where the compost bacteria reduce sulfate and the limestone increases alkalinity. The remediated water is directed through the drainage pipes to the receiving stream. With the use of a Vertical Flow System, waters from the South Fork of the Tangascootack Creek exhibited an increase in pH, from pH 3.6 to pH 7.3, and an increase in alkalinity from 0 mg/l to 130 mg/l (Bisko, 2002).

These passive acid mine drainage treatment systems demonstrate the value of alkalinity in controlling the pH of natural and constructed systems. Similar systems are being used throughout the United States.

**Test Kit Activities**

Perform an alkalinity titration. At least one day prior to the experiment, prepare three beakers: one with tap water, one with a piece of granite or another igneous rock in tap water, and one with limestone, concrete, or an antacid tablet in tap water. As a class, measure and record the alkalinity of each. Add acid, such as vinegar, to each mixture and record the amount of acid as you go. Using a test kit to measure pH, add enough acid to bring each mixture to a pH of 4.2. At this pH, all of the alkalinity is consumed. Does the amount of acid added differ among mixtures?

**References**


Bisko, D. 2002. South Fork Tangascootack Watershed Rehabilitation Project Data. E-mail to the author. March 25, 2002.

Bisko, D. 1998. Tangascootack Creek Watershed Restoration Plan. E-mail to the author. March 22, 2002.

Drever, J. 1997. The Geochemistry of Natural Waters, 3rd Edition. Upper Saddle River, NJ: Prentice Hall.

Hem, J. 1970. Study and Interpretation of the Chemical Characteristics of Natural Water, 2nd Edition. Washington, D.C.: U.S. Government Printing Office.

Kentucky River Assessment Monitoring Project (KRAMP). n.d. Alkalinity and Water Quality. Retrieved on October 2, 2001, from the website: http://water.nr.state.ky.us/ww/ramp/rmalk.htm

Pennsylvania Department of Environmental Protection. n.d. The Science of Acid Mine Drainage and Passive Treatment. Retrieved on March 18, 2002, from the website: http://www.dep.state.pa.us/dep/deputate/minres/bamr/amd/science_of_amd.htm


**Notes:**

# Bacteria

#### Primary Importance

"Bacteria are among the simplest, smallest, and most abundant organisms on earth" (Bartenhagen et al., 1995). Total coliform are a group of bacteria used to indicate the potential presence of harmful bacteria in water.

#### Technical Overview And History

All bacteria are single-celled (unicellular) organisms that do not have a nucleus enclosed in a nuclear membrane, which makes them prokaryotes (Bartenhagen et al., 1995). Only bacteria and blue-green algae (archaea) are prokaryotes; all other life forms have cells containing a nucleus, and are thus eukaryotes.

Within their single cell, most bacteria contain all of the genetic information (DNA) and necessary tools (rhibosomes, protein, etc.) for reproduction via binary fission. In binary fission a single cell divides to form two daughter cells, which are clones of the parent bacterium. Each daughter then divides to form two new daughter cells, which then divide to make four, which then divide to make eight, which then divide to make sixteen, etc. Binary fission can cause one bacterium to "become a billion bacteria in just 10 hours" (American Society for Microbiology, 1999). Thus, bacterial populations grow exponentially until they run out of food, space, or optimal conditions.

**Sources**

Natural
- Bacteria are everywhere

Human
- Human feces–E. coli

**Relevance**

Human
- Indicator of fecal contamination
- Coexists with more harmful bacteria
- Produce useful vitamins

Animals
- Indicator of high bacteria counts
- Produce useful vitamins

**Impacts**

Human health
- Some produce toxins
- Immersion not advisable in waters with high counts
- Can cause illness

Environmental health
- High levels cause recreational water closures

**Possible Remedies**
- Animal waste lagoon
- Stream bank fencing
- Filtration
- Disinfection

Bacteria use sunlight, sugar and starch, sulfur, or iron for food. Most of this food is absorbed from the material in which they live. They are found in homes, restaurants, businesses, soil, water, air, arctic ice, volcanic vents, in and on humans and animals, and everywhere in between. Some use rotating tail-like flagella to navigate through their environment, while others secrete a slime layer that allows them to move (American Society for Microbiology, 1999).

Bacteria range in size from about 0.1 micrometers (µm) in diameter to about 10 µm. They are found in three shapes: rod- or stick-shaped bacteria are called bacilli, spherical or spheroid bacteria are called cocci, and spiral- or corkscrew-shaped bacteria are called spirilla. Bacteria are found individually or in pairs, chains, squares, or other groupings.

Bacteria can be further classified by the degree of oxygen that they prefer (Bartenhagen et al., 1995):
- Obligate (strict) aerobes: require the presence of oxygen (e.g., Pseudomanas fluorescens)
- Obligate (strict) anaerobes: require the absence of oxygen; oxygen is toxic to the cells (e.g., Clostridium botulinum)
- Facultative anaerobes: can survive with or without oxygen (e.g., Escherichia coli)
- Microaerophiles: require a low concentration of oxygen; do not survive in atmospheric oxygen

levels or in environments without oxygen (e.g., Sphaerotilus natans)

Bacterial growth rates are optimized in environments of certain temperature ranges, since they cannot control their own temperature. Classifications based on temperature are as follows (Bartenhagen et al., 1995):

- Psychrophiles: optimum temperature is between 32 and 68°F (0 and 20°C)
- Mesophiles: optimum tem-perature is between 68 and 113°F (20 and 45°C)
- Thermophiles: optimum tem-perature is between 113 and 140°F (45 and 60°C)

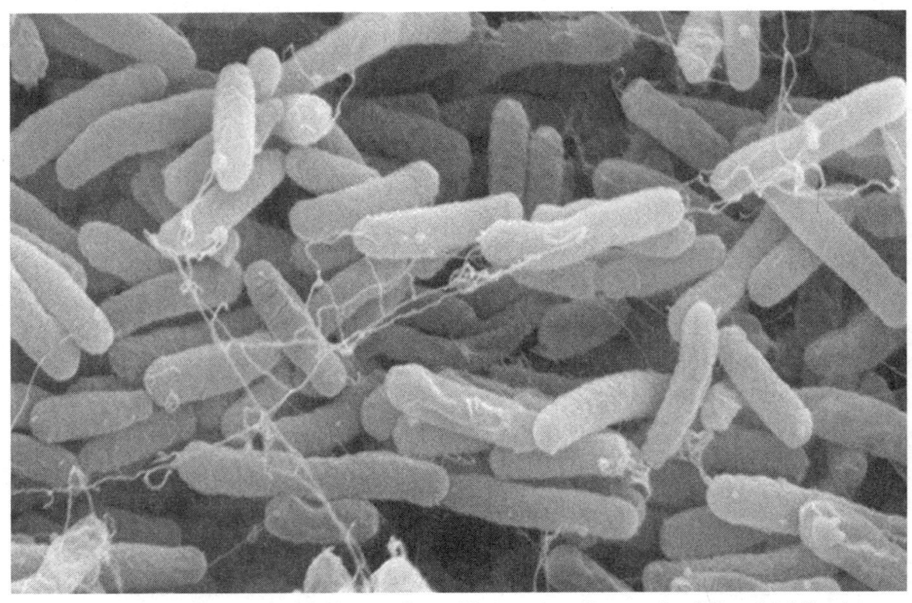

*E. coli bacteria. Courtesy: Michigan State University; Center for Microbial Ecology*

While people often associate bacteria with a lack of cleanliness, many bacteria are beneficial if not necessary for humans, plants and animals survival. Bacteria aide in digestion, help plants grow, break down garbage, and are used to make antibiotics to fight infection. In the environment, bacteria are used to clean up oil spills, hazardous waste dumps and landfills. They are used in the production of beer, wine, cheese, yogurt, chewing gum, paper, and soy sauce to name a few.

Unfortunately, some bacteria produce toxins or cause disease (pathogenic). Many such bacteria are transmitted by water, and therefore may contaminate our drinking water supply. Their presence in water is undetectable by our unaided eyes, nose, or mouth. In order to confirm or deny the presence of these harmful organisms in water, tests must be performed. For municipal water systems, the Environmental Protection Agency (EPA) has established a limit of zero bacteria in drinking water, and therefore water treatment plants must test for them regularly.

"Since 1880, coliform bacteria have been used to assess the quality of water and the likelihood of pathogens being present. Although several of the coliform bacteria are not usually pathogenic themselves, they serve as an indicator of potential bacterial pathogen contamination. It is generally much simpler, quicker, and safer to analyze for these organisms than for the individual pathogens that may be present" (Bartenhagen et al., 1995).

Total coliforms are a group of aerobic and facultatively anaerobic bacteria that include the Genus: Escherichia, Citrobacter, Klebsiella, and Enterobacter. They are ubiquitous in nature. The EPA requires that municipal water treatment facilities test for total coliform bacteria. If a sample is found to be total coliform-positive, the facility must test the same sample for fecal coliforms.

Fecal coliforms are a sub-group of total coliforms, and are distinguished from total coliforms by their ability to withstand higher temperatures (112°F, 44.5°C). Fecal coliforms are found in the intestinal tract of warm-blooded humans and animals, and are therefore transmitted to water and soil by human and animal feces. The average human, whether healthy or ill, excretes billions of fecal coliforms each day through waste (Bartenhagen, et al., 1995). The presence of fecal coliforms in water indicates contamination by

human or animal feces, which can carry other pathogenic bacteria as well.

Escherichia coli (E. coli) are a sub-group of fecal coliforms. Just as humans belong to the Genus Homo and the species sapiens, these bacteria belong to the Genus Escherichia and the species coli. As there are many different individuals in the species sapiens, so too are there many different individuals in the species coli, which are referred to as unique strains. Most strains of *E. coli* reside in our intestines, and produce important vitamins for us, including vitamin K and B-complex vitamins. But some strains of *E. coli* can produce toxins in our intestines, causing symptoms ranging from mild diarrhea to severe intestinal bleeding or hemorrhaging (Brown, 1997).

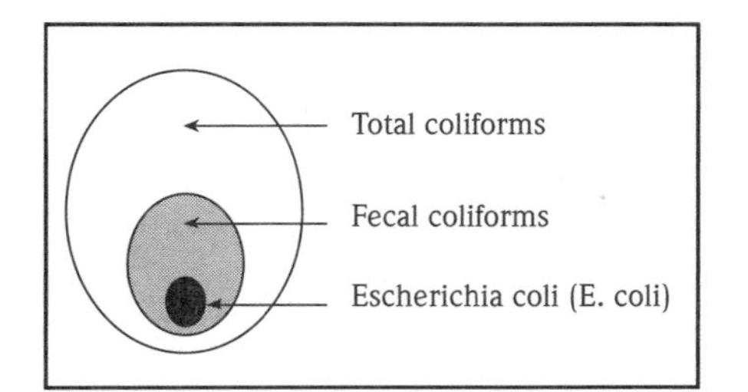


*Graphic representation of the relationship between these coliforms (Lawrence, 2002)*

*Oyster and clam beds are closed due to high levels of fecal coliform bacteria, an indicator of sewage pollution. Courtesy: National Oceanographic and Atmospheric Administration.*

### Sources

When fecal matter from warm-blooded animals comes in contact with water, there is a potential for bacterial contamination. Bacteria can enter water directly from point sources and indirectly from non-point sources. Point sources include leaking sewers, malfunctioning waste water treatment equipment, and sanitary sewer overflows. Nonpoint sources are often carried in runoff, and may include pet and livestock waste, combined sewer overflows, and wildlife waste.

During large runoff events, such as those associated with spring snowmelt, fecal coliform counts are often elevated as runoff water flushes feces from both point and nonpoint sources.

### Relevance

"The Environmental Protection Agency (EPA) requires all public water systems to monitor for total coliforms in distribution systems" (EPA, 1999). Coliform bacteria grow in the intestinal tracts of warm-blooded animals, and are excreted in their feces.

Therefore, these bacteria are monitored as an indicator of potential fecal contamination of water sources. In addition, they are relatively easy to test for, and they are often found in much higher numbers than other bacteria. Coliform bacteria are not necessarily harmful, however feces can carry other pathogenic bacteria that must be removed from drinking and recreational water sources.

### Impacts

E. coli bacteria can contaminate drinking water, ice, recreational waters, under-cooked meat and vegetables, and under-pasteurized juices and dairy products. Fruits and vegetables can carry E. coli when

irrigated with contaminated water. Meat becomes contaminated from improper processing and handling. Contaminated foods and beverages can lead to diarrhea, intestinal bleeding, kidney failure and eye, ear, nose, and throat infections (Bartenhagen et al., 1995).

These harmful E. coli produce toxins in the body. Some of these toxins cause the intestines to produce excessive fluids thus leading to diarrhea. Others, such as the Vero toxin produced by the E. coli O157:H7, cause severe bleeding or hemorrhaging of the intestine and bloody diarrhea.

### Possible Remedies

Preventing human and animal feces from contacting water sources is the best way to avoid bacterial contamination. Several Best Management Practices (BMPs) were developed to mitigate runoff contaminated with bacteria from rangelands, feedlots, and manure storage areas. For example, animal waste lagoons contain wastes from runoff. In addition, stream bank fencing prevents livestock from depositing feces directly in rivers.

When bacteria are already contaminating water, they can be removed via filtration and disinfection. Filtering the water does not necessarily remove bacteria, but prepares the water for effective disinfection by removing debris that can harbor bacteria. Water can then be disinfected using chlorine, iodine, ozone, ultraviolet light, and physical methods such as boiling or sterilization (Oram, n.d.)
In case of an emergency, water can be disinfected as follows:

1. Find the percentage of available chlorine from the label on household bleach.
2. Use the following table, prepared by the Environmental Protection Agency, to determine how much bleach to add to the water:

| Available Chlorine | Drops per Quart of Clear Water |
|---|---|
| 1% | 10 |
| 4-6% | 2 |
| 7-10% | 1 |

(If strength is unknown, add ten drops per quart of water. Double the amount of chlorine for cloudy or colored water)

3. Mix thoroughly and allow the water to stand for 30 minutes
4. The water should have a slight chlorine odor. If not, repeat the dosage and allow the water to stand for an additional 15 minutes. If the water has too strong a chlorine taste, allow the water to stand exposed to the air for a few hours or pour it from one clean container to another clean container several times.

### Case Study

The Chattahoochee River begins in northern Georgia, in the Georgia Mountains of Union County. The name Chattahoochee comes from the Creek Indians, and means river of painted rocks (Lawrence, 2002). It flows south through metropolitan Atlanta, where it provides 70% of the drinking water supply at a rate of more than 300 million gallons per day (Lawrence, 2002). It is also used for power generation, irrigation, industrial supply, and recreation. Water is discharged into the river from urban runoff, industrial effluent, agricultural runoff, and more than 250 million gallons each day of treated waste water from Metropolitan Atlanta.

In 1978, Congress recognized the river as a valuable resource and designated a 48 mile stretch just north of Metropolitan Atlanta as the Chattahoochee River National Recreation Area (CRNRA). "The CRNRA contains about three-fourths of all public green space in a 10-county area of Metropolitan Atlanta, Georgia (Lawrence, 2002). Approximately 3.5 million people visit the CRNRA every year. Many of these visitors are fishing, swimming, and floating in the waters of the Chattahoochee.

With such a large population base surrounding and using the Chattahoochee River, pollution has been a growing concern. Runoff from urban areas carries pet waste (bacteria), oil and gas; agricultural runoff carries sediment, nutrients, animal wastes (bacteria), and pesticides; wastewater carries heavy metals and potentially bacteria.

Fecal coliform bacteria levels have been exceeding Environmental Protection Agency standards for recreational waters around Atlanta since at least 1994, when the United States Geological Survey (USGS) began monitoring for coliform bacteria at Johnson Ferry Road, Atlanta. And this condition is not unique to Atlanta; "bacterial contamination was ranked as the third most common cause for water-body impairment in the United States during 1996" (Armitage and others, 1999, from Gregory et al., 2000).

Fecal coliform bacteria are found in the intestinal tract of warm-blooded animals and enter the water through various point and nonpoint sources. Point sources include leaking sewers, malfunctioning waste water treatment equipment, and sanitary sewer overflows. Non-point sources are often carried in runoff, and may include pet and livestock waste, combined sewer overflows, and wildlife waste.

Several organizations are making efforts to protect the waters of the Chattahoochee, including Upper Chattahoochee Riverkeepers, Trust for Public Land, and the Nature Conservancy of Georgia. One such project, sponsored by the Trust for Public Land (TPL) and The Nature Conservancy of Georgia, is called The Chattahoochee Riverway. Through conservation easements, donations, and land purchases, the organizations hope to secure 500 feet of park land on both sides of the river for 180 miles along the river. This greenway is just a small part of their overall goal to "assure safe, clean drinking water; enhance quality of life in communities along the river; protect an oasis of natural habitat in the midst of a rapidly developing region; and provide places for people to enjoy the river in their daily lives" (Holliday, n.d.). In addition, stream-bank vegetation has been proven to reduce runoff into rivers, which will reduce the input of bacteria, sediment, pesticides, and other constituents carried in runoff (See Turbidity).

In order to monitor their efforts and inform the public of bacterial contamination, the Chattahoochee Riverway Project has also established a bacteriAlert network. State and federal agencies and non-governmental organizations work together

*The Chattahoochee River is enjoyed by millions of recreationalists every year. "Evening at Horseshoe Rapid" Courtesy: Joe and Monica Cook.*

to monitor total coliform and E. coli levels, turbidity, temperature, pH and other parameters daily at three locations along the Chattahoochee: Medlock Bridge near Norcross (north of Atlanta), Johnson Ferry Road near Atlanta, and Paces Ferry Road at Atlanta. The water quality data are then placed on a publicly accessible website within 24 hours (http://ga2.er.usgs.gov/bacteria/default.cfm), emailed to interested parties, and posted at several park entrances.

Scientists are also analyzing the daily monitoring data in an effort to predict bacteria counts based on turbidity levels in the future. Much of the bacteria are carried to the river via runoff, especially during heavy rainstorms. Turbidity levels also tend to increase during heavy rain, therefore leading scientists to try to quantify the relationship. For example, the EPA E. coli recreational water standard was exceeded in all samples at Norcross that exceeded 50 Nephelometric Turbidity Units (NTU), and in all samples at the Atlanta side that exceeded 60 NTU.

By informing the public and preserving and protecting land adjacent to the Chattahoochee, these organizations hope to reduce the amount of bacteria and other pollutants entering the river. They hope to inspire the river's users to become its greatest advocates; those who fish, swim, boat or meander along the banks of the Chattahoochee can make a difference in the future of their river.

**Test Kit Activities**

Identify potential sources of fecal coliform contamination in water, such as runoff from urban parks, feedlots, rangelands, pastures, croplands fertilized with manure, and a wastewater treatment plant. Collect samples from as many sources as possible during the dry season, when waters are low, and test for fecal coliforms. Collect samples from the same locations during the high-water season. Compare the bacterial levels between seasons. Why might they differ? Be sure students wear gloves and wash hands thoroughly when working around potentially contaminated water sources.

**References**


American Society for Microbiology (ASM). 1999. Microbial Reproduction. Retrieved on January 28, 2002, from the website: http://www.microbe.org/microbes/reproduction.asp

Bartenhagen, K., M. Turner, and D. Osmond, 1995. Bacteria. Retrieved on January 15, 2002 from the North Carolina State University WATERSHEDDS: Water, Soil and Hydro Environmental Decision Support System web site: http://h2osparc.wq.ncsu.edu/info/bacteria.html

Brown, J. 1997. What the Heck is an E. coli? Retrieved on May 14, 2002, from the University of Kansas Department of Molecular Biosciences website: http://people.ku.edu/%7Ejbrown/ecoli.html

Gregory, B. and E. Frick, 2002. Fecal-coliform bacteria concentrations in streams of the Chattahoochee River National Recreation Area, Metropolitan Atlanta, Georgia, May-October 1994 and 1995. Retrieved on May 20, 2002, from the United States Geological Survey online Water-Resources Investigations Report 00-4139: http://wwwga.usgs.gov/publications/wrir00-4139.pdf

Holliday, P. n.d. The Chattahoochee: A River Facing Peril and Possibility. Retrieved on May 20, 2002, from the Sherpa Guides website: http://sherpaguides.com/georgia/chattahoochee/chattahoochee_in_peril/index.html

Lawrence, S. 2002. Chattahoochee Riverway Project. Retrieved on May 9, 2002, from the Chattahoochee Riverway Project website: http://ga2.er.usgs.gov/bacteria/SummaryIntroduction.cfm

Environmental Protection Agency (EPA). 1999. Drinking Water Pathogens and Their Indicators: A Reference Resource. Retrieved on May 9, 2002, from the website: http://www.epa.gov/enviro/html/icr/gloss_path.html

Oram, B. n.d. Total Coliform Bacteria. Retrieved on January 15, 2002, from the Wilkes University Center for Environmental Quality website: http://wilkes.edu/~eqc/coliform.htm

# Conductivity

## Primary Importance

Conductivity is a quick and easy estimate of the total dissolved solids in a sample, which can serve as a warning for more serious water quality problems.

## Technical Overview and History

Conductivity, also known as specific conductance, is a measure of how well a water sample conducts electricity. The reciprocal of conductivity is resistivity, a measure of how resistive water is to conducting an electrical current. The less resistance the water has, the higher its conductivity.

The presence of ions in water makes it a good conductor of electricity. Ions are atoms or molecules that become charged when they either gain or lose electrons. When an atom gains an electron it becomes a negatively charged ion, or an anion. When an atom loses an electron it becomes a positively charged ion, or a cation.

Ions that are often found in natural waters include: calcium ($Ca^{2+}$), aluminum ($Al^{3+}$), magnesium ($Mg^{2+}$), sodium ($Na^{+}$), potassium ($K^{+}$), carbonate ($CO_3^{2-}$), bicarbonate ($HCO_3^{-}$), phosphate ($PO_4^{3-}$), chloride ($Cl^{-}$), nitrate ($NO_3^{-}$), and sulfate ($SO_4^{2-}$). The presence of these ions increases the conductivity of water. Many of these ions are useful and, in fact, necessary. Nutrients such as nitrates and phosphates are required for plant and animal growth. Organisms often need calcium and magnesium to maintain bones, shells and teeth.

### Sources

Natural
- Geology
- Soil/sediment

Human
- Mining
- Chemical de-icers/ road salt
- Industrial discharge
- Sewage effluent
- Poor irrigation practices

### Uses

Human
- Pre-screen for water quality
- Trace water movement

Animals and Plants
- Essential ingredients for aquatic life

### Impacts

Human health
- Potential chronic illness

Environmental health
- Wilting, leaf drop, root death for plants
- Hazardous chemicals affect all organisms

### Possible Remedies

- Proper irrigation techniques
- Minimize road salt, chemical de-icers
- Effective treatment of sewage and industrial effluent
- Desalination

Other materials such as silica ($SiO_2$), sugar, oils, alcohols, and many hazardous organic compounds (e.g., benzene) do not affect conductivity because they do not ionize, (yield charged molecules or atoms) when dissolved in water.

Conductivity is measured by using a specially designed voltmeter and conductivity probe. A conductivity probe has two electrodes separated by a known distance (usually 1 cm). When the probe is immersed in the water sample, a voltage is applied across the electrodes and the voltage drop is measured. A small drop in voltage indicates that the water has a high conductivity. A large drop in voltage, such as that seen in a sample of pure, deionized or distilled water, indicates that the water has a low conductivity. It is important to note that the conductivity measurement does not identify individual ions, it only indicates whether there are many or few ions in the sample. Conductivity is commonly measured in microsiemens/cm (µS/cm) or, as the reciprocal of resistance (measured in ohms), as mhos/cm.

The temperature of the water affects the conductivity measurement. As the water temperature increases, ions in solution move more rapidly, and allow the water to conduct electricity better, yielding an increased conductivity value. Therefore, the temperature

must be taken and recorded whenever measuring conductivity. Many conductivity probes have a built-in temperature sensor, so the associated conductivity meter can automatically compensate for the effect of temperature. Although a variety of temperature adjustment techniques exist, often conductivity is standardized to 25° C by either increasing "the reading by 2% for every degree below 25° C, [or decreasing] the reading by 2% for every degree above 25° C", (Melbourne Parks & Waterways, 1995).

| Average Conductivity Levels | |
|---|---|
| Deionized water: | 0–1 µS/cm |
| Healthy stream: | 150–500 µS/cm |
| Seawater: | 50,000 µS/cm |

**Sources**

The geology of an area greatly influences the type and amount of ions in its water bodies. For example, rivers that run through granite bedrock tend to be clear with low conductivity. Their waters have few ions because granite weathering yields primarily silica ($SiO_2$), which is not an ion. Areas on the other hand containing limestone or clays tend to support waters with high conductivity because these materials dissolve easily to release charged ions like calcium, magnesium, and carbonate into the water.

Anthropogenic (human-caused) sources of ions in waterways include mining operations, agriculture, sewage effluent, and some industrial discharges. Mining operations can add iron, sulfate, copper, cadmium, and arsenic ions. Agriculture irrigation can add nutrient ions such as nitrates and phosphates to the water. Irrigation can also leach salts from soil and carry them into nearby waters. Sewage effluent can contain chloride, nitrates, and phosphates, all ions that increase conductivity. Urban runoff carries auto fluids, salts and chemical de-icers.

While some industrial effluents increase conductivity, others decrease conductivity. Effluent containing salt, metals, chloride, or other ions increases the conductivity of the body of water into which it is discharged. Effluent containing oils, alcohols, phenols or other non-charged ions tend to decrease the conductivity by reducing the number of charged ions per unit volume of water.

**Relevance**

Conductivity is measured for several reasons. Because it increases when ions increase, conductivity can be a quick and effective estimate of the total dissolved solids (TDS) in a sample (Melbourne Parks & Waterways, 1995). Measuring TDS directly requires a more tedious, gravimetric technique. Conductivity also serves as an immediate indicator of problems in wastewater treatment or natural systems. Farmers can use conductivity as a quick and easy estimate of the salinity of irrigation water or irrigation return flow. Though conductivity is a useful measure and can point to potential problems that may require further investigation; it alone cannot diagnose, create, or remedy a water quality problem.

Environmental engineers or limnologists may also use conductivity to trace water movement. Using conductivity, they can track the path a tributary or an industrial effluent follows within a water column. Since ions are relatively unreactive once they are in the water, if the conductivity of the influent water differs from the conductivity of the water into which it flows, that conductivity difference can be mapped to trace the path of the tributary water.

Employing a similar methodology, conductivity can also measure the speed with which water flows through a water body. First, water at the inlet is spiked with a large amount of an unreactive ion that is not typically found in the system (for example, bromide). Then, the conductivity of water at the outlet is measured until the spike is observed. The residence time of water in a body, such as a pond or wetland, can be calculated with this information.

**Impacts**

While a high or low conductivity measurement cannot identify which ions are out of balance, it can indicate the need for more specific water quality tests to pinpoint the water quality problem. Arsenic, for example, is a human carcinogen according to the International Agency for Research on Cancer (IARC)

(Galal-Gorchev, 2000). Too much calcium or iron may cause build up in water pipes or discoloration in sinks and bathtubs, but poses no health threat. Excessive salts may be harmful to plants, causing wilting, root death, and leaf drop.

At the same time, too few ions can be harmful. Waters with low conductivity due to high levels of oils or hydrocarbons are not healthy for the aquatic organisms or humans that come in contact with them. Additionally, some ions serve as food sources for aquatic organisms and their absence poses a survival risk to those organisms.

Waters with very low conductivity also tend to be more corrosive. When very low conductivity waters flow through pipes, they have a greater tendency to leach out metals such as lead and copper than do waters with moderate conductivity (Hach Company, 1997). Often, chemicals will be added to drinking water to increase its conductivity and reduce its potential for corrosiveness.

### Possible Remedies

While little can be done to control natural sources of conductivity, proper management of all anthropogenic sources of ions can remedy virtually any ion imbalance in waterways. Proper management includes irrigation practices that limit salt leaching and erosion, successful treatment of sewage effluent, and proper treatment of industrial effluent. Once salts are in the water, the water may need to be treated to lower its conductivity. Water treatment methods to decrease conductivity include dilution, ion exchange, and reverse osmosis.

### Case Study

One of the major environmental issues facing the Colorado River is its increasing level of total dissolved solids, mostly in the form of dissolved salts or salinity. Near its headwaters, the salinity levels of the Colorado River are relatively low. The igneous and metamorphic rocks of Rocky Mountain National Park, where the river originates, do not erode easily and contribute very little salinity to the river. According to the United States Geological Survey, salinity near the headwaters is generally less than 50 mg/l (ppm) (Hart, et. al, 2000).

As the river flows to the Gulf of Mexico, the salinity concentration increases. Roughly half of the salinity added to the Colorado River comes from natural sources, including the geology, evaporation, and influent springs (Pontius, 1997). Downstream from the headwaters, the river eventually flows over sedimentary rocks. The action of water flowing over these rocks causes them to erode, dissolving salts and picking up sediment, thus increasing the total dissolved solids or conductivity. Since the river flows through an arid region, water evaporates from its surface and the volume of water is decreased. This evaporation increases the concentration of salt and sediments in the water. Finally, mineral springs in the Colorado River watershed contribute salts and other minerals to its waters. In fact, "natural hot springs contribute about 500,000 tons of dissolved solids annually to the streams in the basin" (Apodaca, et. al, 1996, p. 2).

The other half of the Colorado River salinity comes from human sources, such as municipal and industrial effluents, reservoir evaporation, phreatophyte (water-loving plants, such as willows) use, and irrigation return flow.

High salinity levels corrode pipes by disrupting protective scale formation and increasing the cost of drinking water treatment. Waters with high salinity used for irrigation decrease crop yields and require greater volumes of water to dilute the build-up of salts (Pontius, 1997).

*Lining a canal with cement to reduce salinity and conductivity by controlling seepage. Courtesy: Gomaco Corporation.*

For example, "In 1964, salinity became an international issue when the Mexican government complained that deliveries of Colorado River water with salt concentrations of 2,000 ppm were affecting [farmers'] ability to grow crops and asserted that this was in violation of the 1944 Mexican Water Treaty" (Pontius, 1997, p. 60). This Treaty secured a quantity of water for Mexico from the Colorado River, but did not address water quality. These assertions precipitated a dialogue that would eventuate in Minute No. 242 to the Treaty.

Minute 242 of the Mexican Water Treaty was signed in 1974 and the Colorado River Basin Salinity Control Act (CRBSCA) was passed to implement the Treaty. Both Title I and Title II of the CRBSCA authorize measures to control the salinity of the Colorado River. Under Title I, construction projects initiated included the Yuma Desalting Plant, a canal to bypass nearby irrigation waters, and a well field to supplement river flows. In addition, pumping of water near the border was restricted, 10,000 acres of farmland was retired, and canals were lined to reduce seepage. Title II designated four salinity control units, or regions, within the Colorado River Basin.

The basin-wide research program and subsequent salinity control measures have been so successful that the Yuma Desalting Plant is no longer in operation. While Minute 242 and the CRBSCA were precipitated by salinity levels of 2,000 ppm, salinity levels at the border between the United States and Mexico now are about 900 ppm (Hart, et. al, 2000). Additional salinity control research and implementation has been, and will continue to be, conducted in the Colorado River watershed.

**Test Kit Activities**

Collect soil samples from various locations, such as agricultural land, wetlands, forest, garden, playground, etc. Break up large chunks, and spread soil on a table, desk, or countertop to air dry. Label samples, and prevent cross-contamination by separating them. Drying should take several days, depending on humidity levels and moisture content of the soil.

Next, place about 20 grams of dry soil from each sample in separate beakers and label the beakers accordingly. Add 20 ml of deionized water to each beaker. Stir soil and water samples for one minute at 10-minute intervals over a thirty minute period. Measure the conductivity of each sample. This is an approximation of the soil salinity. Multiply the conductivity values by 0.67 for an estimate of the total dissolved solids in the soil water.

**References**


Apodaca, L., V. Stephens, and N. Driver. 1996. *What Affects Water Quality in the Upper Colorado River Basin.* Retrieved on April 19, 2002, from the United States Geological Survey website: http://webserver.cr.usgs.gov/Pubs/fs/fs109-96/pdf/fs109-96.pdf

Galal-Gorchev, H. 2000. *Hazardous Chemical in Human and Environmental Health.* Retrieved on December 10, 2001 from the World Health Organization International Programme on Chemical Safety web site: .http://www. who.int/pcs/training_material/hazardous_chemicals/section_1.htm

Hach Company. 1997. *Water Analysis Handbook, 3rd Edition.* Loveland, CO: Hach Company.

Hart, R. and R. Hooper. 2000. *Monitoring the Quality of the Nation's Largest Rivers.* Retrieved on April 19, 2002, from the United States Geological Survey website: http://water.usgs.gov/nasqan/progdocs/factsheets/clrdfact/clrdfact.html

Pontius, D. 1997. *Colorado River Basin Study.* Retrieved on April 15, 2002, from the Water in the West website: http://www.waterinthewest.org/reading/readingfiles/fedreportfiles. colorado.pdf

R. Wetzel, 2001. *Limnology, 3rd Edition.* San Diego, CA: Academic Press.

1995. *Physical and Chemical Tests: Conductivity.* Retrieved on December 5, 2001 from the Melbourne Parks and Waterways web site: http://redtail.eou.edu/streamwatch/swm18.html

# Dissolved Oxygen
## and Biochemical Oxygen Demand

### Primary Importance

Just as animals need oxygen to live, aquatic animals need oxygen dissolved in water.

### Technical Overview And History

Dissolved oxygen (DO) is, simply, oxygen that is dissolved in water. Measuring DO in water indicates how much DO is present but not how much oxygen the water is capable of dissolving. "When water dissolves all of the oxygen it is capable of holding at a given temperature it is said to be 100 percent saturated" (Florida Oceanographic Society, 2002).

A dissolved oxygen concentration can be converted to a percent saturation value by using the following table adapted from the Streamkeeper's Field Guide (Murdoch, et. al, 1996).

1. In the field, measure and record the temperature of your water sample.
2. To calculate the percent saturation, divide the dissolved oxygen concentration of your sample by the maximum concentration (100% saturation) at the temperature of your sample, listed on the next page: Note: these values may also be adjusted for altitude. See Mitchell et al., 1997, p. 21.

### Sources

Natural

- Absorption at air-water interface
- By-product of plant photosynthesis

### Relevance

Animals

- Essential for the survival of most aquatic animals
- Fish absorb through gills
- Used for respiration

### Impacts

Human Health

- No human health impacts

Environmental health

- Low DO causes species shifts
- Potential for fish kills

### Possible Remedies

- Maintain natural water temperature
- Limit erosion
- Limit human sources of organic matter

While there are many factors that influence dissolved oxygen (see Sources), one that significantly impacts percent saturation is the amount of organic matter present in water. Organic matter–debris that comes from living or once-living plants and animals–is decomposed by microorganisms that consume oxygen in this process. The more organic matter present, the more oxygen is consumed. The amount of oxygen consumed is called the biochemical oxygen demand (BOD) (also referred to as the biological oxygen demand). Biochemical oxygen demand also measures the oxygen removed from water during chemical reactions, such as the oxidation of sulfides, ferrous iron, and ammonia (Bartenhagen, et. al, 1995).

Dissolved oxygen can be consumed quickly in waters with a high biochemical oxygen demand leaving no oxygen available for fish. Organic matter that contributes to a high BOD includes dead plants and animals, animal and pet waste, leaves and woody debris, fertilizers, urban runoff, agricultural runoff, effluent, and wastewater treatment plant effluent.

### Sources

Oxygen is dissolved in water either directly at the air-water interface, or as a byproduct of plant photosynthesis. Dissolved oxygen comes out of solution and is removed from water when the water temperature increases. It is also consumed during the respiration of animals and the decomposition of

organic matter by microorganisms (Bartenhagen, et. al, 1995).

The amount of oxygen dissolved in water at any given time depends on several factors, including the water temperature, turbidity, rate of photosynthesis, volume of organic matter, amount of dissolved solids, degree of agitation, source of the water, and prevailing barometric pressure.

The colder the water temperature, the more oxygen can be dissolved in it. Therefore, as water temperature rises, (see Temperature for causes), dissolved oxygen levels decrease.

Turbidity indirectly influences dissolved oxygen levels in two ways. Suspended solids that cause turbidity absorb solar radiation which, in turn, increases the water temperature. Highly turbid water also decreases the amount of penetrating sunlight which decreases photosynthesis and its by-product, oxygen. Non-photosynthesizing plants will die more quickly and contribute to the organic matter that feeds oxygen-consuming bacteria and other microorganisms.

Water with lower dissolved or suspended solids tends to have a higher level of dissolved oxygen than water with higher dissolved or suspended solids. For example, fresh water dissolves more oxygen than salt water.

Agitating water at the surface mixes atmospheric oxygen with the water. When agitation stirs the water enough to replace the oxygen-rich surface water with deeper water, even more oxygen can be dissolved

Maximum Dissolved Oxygen Concentration (100 percent saturation)

| Temperature °C | Dissolved Oxygen mg/L | Temperature °C | Dissolved Oxygen mg/L | Temperature °C | Dissolved Oxygen mg/L |
|---|---|---|---|---|---|
| 1 | 14.60 | 16 | 10.07 | 31 | 7.41 |
| 2 | 14.19 | 17 | 9.85 | 32 | 7.28 |
| 3 | 13.81 | 18 | 9.65 | 33 | 7.16 |
| 4 | 13.44 | 19 | 9.45 | 34 | 7.05 |
| 5 | 13.09 | 20 | 9.26 | 35 | 6.93 |
| 6 | 12.75 | 21 | 9.07 | 36 | 6.82 |
| 7 | 12.43 | 22 | 8.90 | 37 | 6.71 |
| 8 | 12.12 | 23 | 8.72 | 38 | 6.61 |
| 9 | 11.83 | 24 | 8.56 | 39 | 6.51 |
| 10 | 11.55 | 25 | 8.24 | 40 | 6.41 |
| 11 | 11.27 | 26 | 8.09 | 41 | 6.31 |
| 12 | 11.01 | 27 | 7.95 | 42 | 6.22 |
| 13 | 10.76 | 28 | 7.81 | 43 | 6.13 |
| 14 | 10.52 | 29 | 7.67 | 44 | 6.04 |
| 15 | 10.29 | 30 | 7.54 | 45 | 5.95 |

at the air-water interface. This process of mixing continues to add dissolved oxygen to the water. For this reason, fast-flowing rivers that tumble over rocks and falls typically have higher dissolved oxygen levels than slowly meandering rivers. Because slow-moving or stagnant waters undergo little internal mixing, waters at the surface tend to be higher in dissolved oxygen than the rest of the water column. Two water sources have a particularly strong influence on dissolved oxygen levels. First, ground water tends to be low in dissolved oxygen, because it is not exposed to the atmosphere. Second, reservoir water released from the bottom of a dam tends to be low in dissolved oxygen. Both of these sources supply cold water, and while the immediate impact to a receiving water is to lower overall dissolved oxygen level, cold waters actively begin absorbing oxygen from the atmosphere and eventually help increase the receiving water's overall DO level.

Non-varying factors such as altitude have long term implications for a water's DO content. As atmospheric pressure decreases with an increase in elevation, less dissolved oxygen is absorbed by the water. Therefore, assuming all other factors are equal, lakes and rivers at higher elevation have lower dissolved oxygen levels than those at sea level.

## Relevance

Dissolved oxygen is an important water quality parameter to monitor because it is essential for the survival of aquatic life and the health of the lakes and rivers. Fish absorb oxygen through their gills to "breathe" (Murphy, 2001). Aquatic plants and animals require dissolved oxygen to respire. Microorganisms gain energy from dissolved oxygen to decompose organic matter.

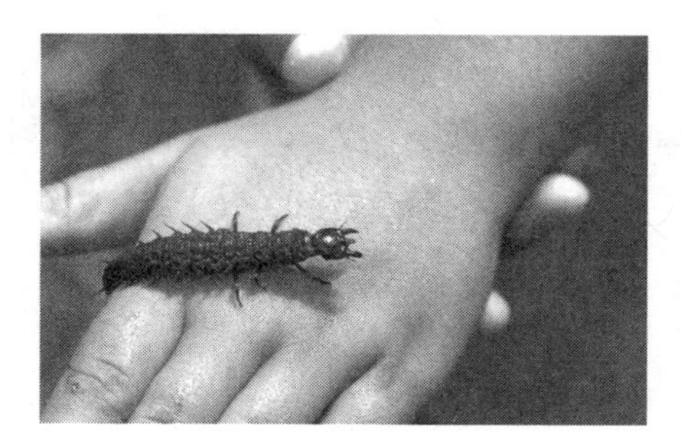

*Helgrammites are affected by changes in dissolved oxygen levels. Courtesy: Hoosier River Watch*

Dissolved oxygen levels less than about 2 mg/l (ppm) cannot support fish and many other aquatic organisms. Water with between 3 and 4 mg/l (ppm) of dissolved oxygen is stressful for most aquatic organisms. More than 5 mg/l (ppm) of dissolved oxygen is required to maintain a healthy and diverse aquatic organism population (Hartman, et. al, 2000).

## Impacts

When dissolved oxygen levels drop below about 4 mg/l (ppm), most aquatic organisms cannot reproduce or grow normally (Hartman, et. al, 2000). Aquatic organisms exposed to dissolved oxygen levels outside of their ideal range become stressed and more susceptible to disease. Even moderately decreased DO levels can ultimately lead to death if the organism does not relocate to waters with higher dissolved oxygen.

As dissolved oxygen drops below about 5 mg/l (ppm), resident species are replaced by pollution-tolerant species. For example, when DO decreases, so will organisms such as trout, pike, stonefly and mayfly nymphs, and caddisfly and beetle larvae. Meanwhile catfish, carp, fly larvae, worms, and leech populations increase because they can survive in waters with low levels of dissolved oxygen. In addition, waters of low dissolved oxygen can support nuisance algae and anaerobic organisms.

Often waters reach their lowest daily level of dissolved oxygen just before dawn because animals have consumed oxygen, and plant photosynthesis slows throughout the night. This phenomenon is compounded in the late summer when water temperatures are at their highest. Therefore, fish kills due to lack of dissolved oxygen are most likely to occur in the early mornings of late summer.

## Possible Remedies

Since there are many factors that influence dissolved oxygen, there are many ways to maintain healthy levels. Human impacts that raise water temperature–such as thermal pollution and the removal of vegetation along stream banks–should be minimized.

Since unnaturally high turbidity increases the water temperature

and decreases plant photosynthesis, erosion should be controlled. Tactics such as maintaining a buffer strip of vegetation along stream banks, planting streamside vegetation where it has been removed, implementing farming Best Management Practices (BMPs) for erosion prevention, and protecting new construction sites from erosion all target this goal.

Naturally occurring levels of organic matter generally do not negatively affect dissolved oxygen levels. However, human sources of organic matter such as pet waste, fertilizers, urban runoff, agricultural runoff, industrial effluent, and waste water treatment plant effluent can contribute excessive organic matter. These unnatural organic matter levels accelerate dissolved oxygen consumption by aerobic bacteria and other microorganisms during the decomposition process. Limiting anthropogenic sources of organic matter can help to maintain a healthy and diverse aquatic organism population.

*Bull Shoals Dam on the White River in Arkansas . Courtesy: the Tennessee Valley Authority.*

### Case Study

Dams provide clean energy as well as recreational opportunities in both the reservoir they create and the cold waters they release. However, these release waters can also be harmful for some native fish populations. Dams that release water from the bottom of the reservoir are of particular concern.

Temperate lakes and reservoirs undergo seasonal stratification due to the density differences in water as it changes temperature. In general warm water is less dense than colder water, so warm water remains on the surface, while colder water sinks. In early spring, most of a temperate water column is the same temperature due to mixing and a phenomenon know as turn-over. When the water is well mixed, dissolved oxygen levels are fairly consistent throughout. However, during the summer the temperature changes as surface water warms, causing plant and animal life to flourish. These plants and animals die and sink to the colder bottom water. Dead plants and animals are decomposed by microorganisms, which consume oxygen in the process.

As microorganisms break down more organic matter (dead plant and animal matter), dissolved oxygen in the bottom waters becomes depleted. By late summer and early fall, dissolved oxygen levels in bottom waters can be too low to support most aquatic life.

This dissolved oxygen-deficient, cold water from the bottom of the reservoir is then released. Warm water species, such as bass and sunfish, are forced from their native water or are killed by the cold release waters. Often, release waters have been stocked with cold water species such as rainbow and cutthroat trout that require higher dissolved oxygen levels to survive.

Low dissolved oxygen levels are suspected to have caused fish kills on the White River in Arkansas and Missouri in 1954, 1963, 1964, 1971, 1972, and 1990 (Langton, 1994). Immediately following the 1990 fish kill, dissolved oxygen levels below the Bull Shoals Dam of the White river were measured to be less than 2 mg/l (ppm), a level that is often lethal for trout.

This fish kill led representatives from the Army Corps of Engineers (Corps), Southwester Power Administration, Arkansas Game and Fish Commission, Arkansas Department of Pollution Control and Ecology, Arkansas Department of Parks and Tourism, and Arkansas Soil and Water Conservation Commission to form what is now known as the White River Dissolved Oxygen Committee. They sought a solution that would protect both the hydroelectric revenue created by the dam and the lucrative trout fishing that had grown to become "one of the single largest revenue producing industries in the state" (Langton, 1994, p. 1).

The committee decided that dissolved oxygen levels below the Bull Shoals and Norfork dams would be monitored by the United States Geological Survey (USGS) and the Corps and the data would be made available online to all interested parties via the Corps computer system.

The committee also examined the successful results of technology applied at other dam operations. In 1993 the Corps installed hub-baffles on the dam turbines (Langton, 1994). The following table of dissolved oxygen levels demonstrates that the hub-baffles are significantly improving dissolved oxygen levels below the dams.These levels are much higher than the 2 mg/l dissolved oxygen levels in 1990 that precipitated the fish kill. It is clear that hydropower from dams and healthy fish populations below the dams can coexist.

| | At Bull Shoals Dam, near Flippin | Below Bull Shoals Dam, at Bull shoals | Below Bull Shoals Dam, near Fairview |
|---|---|---|---|
| | (07054501) | (07054502) | (07054527) |
| Oct., 1995 | 6.30 mg/l DO | 6.91 mg/l DO | 7.09 mg/l DO |
| Oct., 1998 | 4.15 mg/l DO | 8.08 mg/l DO | 8.00 mg/l DO |
| Oct., 2000 | 6.37 mg/l DO | 8.51 mg/l DO | 6.74 mg/l DO |
| Oct., 2001 | 7.48 mg/l DO | 9.87 mg/l DO | 8.21 mg/l DO |

Data provided by the Arkansas Soil and Water Commission, 2002

Table 1: Average dissolved oxygen levels for the month of October in release waters from Bull Shoals Dam, White River, Arkansas in mg/l (ppm)

**Test Kit Activities**

Collect two water samples each, from at least two of the following water sources: lake, river, pond, puddle, and wetland. Measure the dissolved oxygen of one of your samples in the field when you collect the sample. Cap and save the second sample for the biochemical oxygen demand (BOD) test to be performed later. Allow the BOD samples to stand, while covered, for approximately five days. Conduct a DO test on the samples. Compare the differences–this is the BOD for the samples. How do the levels compare?

**References**

Bartenhagen, K, M. Turner, and D. Osmond. 1995. *Dissolved Oxygen.* Retrieved on October 29, 2001 from the North Carolina State University WATERSHEDDS: Water, Soil and Hydro Environmental Decision Support System web site: http://h2osparc.wq.ncsu.edu

Florida Oceanographic Society. 2002. *Dissolved Oxygen.* Retrieved on December 3, 2001, from the website: http://www.fosusa.org/parameters.htm

Hartman, L. and M. Burk. 2000. *Hoosier Riverwatch Guide.* Indianapolis, IN: Natural Resources Education Center.

Langton, S. 1994. *The Experience of the White River Dissolved Oxygen Committee.* Retrieved on March 11, 2002, from the Civil Practices Network website: http://www.cpn.org/cpn/sections/topics/environment/stories-studies/army-corps_langton4b.html

Mitchell, M. and W. Stapp. 1997. *Field Manual for Water Quality Monitoring.* Dubuque, IA: Kendall/Hunt Publishing Company.

Murdoch, T. and M. Cheo. 1996. *The Streamkeeper's Field Guide.* Everette, WA: The Adopt-A-Stream Foundation.

Murphy, S. 2001. *General Information on Dissolved Oxygen.* Retrieved on December 3, 2001, from the Boulder Area Sustainability Information Network website: http://bcn.boulder.co.us/basin/

# Hardness

**Primary Importance**

The water hardness test is significant for residential consumers who find hard water interferes with household cleaning effectiveness. This test is also crucial to industrial processes seriously impacted by the effects of hard water on expensive equipment as well as their final product.

**Technical Overview And History**

Hard water received its name from the fact that it is difficult (hard) to produce a soapy lather using such water. Hardness measures the concentration of dissolved minerals by measuring polyvalent cations, which are ions with a positive charge of two (2) or greater. The most prevalent polyvalent cations that cause hardness are calcium ($Ca^{2+}$) and magnesium ($Mg^{2+}$). Although aluminum ($Al^{3+}$), barium ($Ba^{2+}$), iron ($Fe^{2+}$, $Fe^{3+}$), manganese ($Mn^{2+}$, $Mn^{3+}$), strontium ($Sr^{2+}$), and zinc ($Zn^{2+}$) can also contribute to hardness, calcium and magnesium ions are usually the only ions present in significant concentrations. Therefore, hardness is generally considered to be a measure of the calcium and magnesium content of water (Hach Company, 1997).

Chemical compounds that increase hardness when dissolved in water include: $CaCO_3$ (calcium carbonate), $MgCO_3$ (magnesite), $CaSO_4$ (gypsum), $MgSO_4$ (epsomite), and $CaMg(CO_3)_2$ (dolomite).

Sources

Natural
- Rocks and soil
- Industrial waste

Relevance

Human
- Taste threshold
- May interfere with water treatment
- Removed before used by industries

Animals
- Important components of shells and bones

Plants
- Important components of cell walls
- Component of chlorophyll

Impacts

Human health
- Skin irritation (minor)
- Dull, lifeless hair
- Scale build-up in pipes
- Pipe corrosion

Environmental health
- Decreases metal toxicity in fish/aquatic life
- Some necessary for shell/cuticle secretion
- Maintains osmotic balance in aquatic plants

Possible Remedies
- Use biodegradable surfactants
- Treat water with softener

**Some minerals contributing to hard water (American Chemical Society, 2002, p. 77)**

| Mineral | Chemical formula | Ions from dissolution |
|---|---|---|
| Limestone/ chalk | $CaCO_3$ | $Ca^{2+}$, $CO_3^{2-}$ |
| Magnesite | $MgCO_3$ | $Mg^{2+}$, $CO_3^{2-}$ |
| Gypsum | $CaSO_4 \cdot 2H_2O$ | $Ca^{2+}$, $SO_4^{2-}$ |
| Dolomite | $CaCO_3 \cdot MgCO_3$ | $Ca^{2+}$, $Mg^{2+}$, $CO_3^{2-}$ |

Although concentrations may vary, typical natural waters have concentrations of 4–400 mg/L calcium and 1–1350 mg/L magnesium, (Brunlow, 1996). When measuring hardness, a total hardness value is often reported as mg/L of calcium carbonate ($CaCO_3$). Total hardness includes hardness contributed by both calcium and magnesium ions. Certain tests can specifically measure calcium hardness, which can then be subtracted from the total hardness to determine magnesium hardness.

Drinking water ideally contains between 10 and 500 mg/l of $CaCO_3$ (World Health Organization, 1993). There are no primary or secondary drinking water standards for hardness because it is not a health threat to humans.

| Concentration (mg/l CaCO3) | Classification of water |
|---|---|
| 0 to 60 | soft water |
| 61 to 120 | moderately hard |
| 121 to 180 | hard |
| 180 and greater | very hard |

While classification limits vary slightly among organizations, the United States Geological Survey (Briggs et al., 1977) and the Water Quality Association (2000) define water hardness levels as above:

Water becomes hard when it contacts and then dissolves minerals, metals, salts, and other molecules. It dissolves carbon dioxide ($CO_2$) from the atmosphere to form a weak acid called carbonic acid.

**Carbonic acid:**
$H_2O_{(l)} + CO_{2(g)} = H_2CO_{3(aq)}$

**Carbonic acid dissolves calcium carbonate:**
$CaCO_3 + H_2CO_3 = Ca^{2+} + 2HCO_3^-$ $H\Delta$ rxn = -18KJ/mol

**Calcium carbonate, magnesium carbonate dissolution:**
$CaCO_{3(s)} = Ca^{2+}_{(aq)} + CO_{3\ (aq)}^{2-}$
$MgCO_{3(s)} = Mg^{2+}_{(aq)} + CO_{3\ (aq)}^{2-}$

As this slightly acidic water passes over soil, rock, and organic matter, it dissolves more minerals, metals, salts, and molecules in a process known as weathering.

## Sources

The most significant source of calcium and magnesium ions is the ground. When water passes over certain kinds of bedrock or soil derived from bedrock, it dissolves and carries ions with it.

If the bedrock is limestone, chalk, magnesite, gypsum, or dolomite, the water passing over it will typically contain high concentrations of calcium and magnesium ions. Therefore, areas with an abundance of carbonate rocks (such as limestone) often have hard water. If water passes over granitic or igneous rocks, which do not dissolve easily, there will be few ions in solution. Granite rock primarily contains the minerals quartz, feldspar, and mica. When water flows over these rocks, the minerals break down very slowly into smaller pieces instead of dissolving in water as calcium carbonate does. Hence, areas with primarily igneous rocks usually have soft water.

Because calcium and magnesium also are common constituents in food, we find these ions in food waste and human waste. While most are removed during the wastewater treatment process, some ions remain.

Mining and rock quarry activities contribute to hard water by exposing rocks containing calcium and magnesium. In addition, some cleaning agents and industrial discharge increase the concentration of calcium and magnesium in the environment.

## Relevance

Calcium and magnesium in moderate amounts are necessary for human, animal, and plant life. Aquatic organisms such as crayfish and snails require calcium to produce and maintain their shells or cuticles. Calcium helps aquatic plants and animals maintain a healthy osmotic balance, and plants also need it for their cell walls. Magnesium is necessary for chlorophyll production, making it an essential nutrient for plants. Because very hard water, with 500 mg/l $Ca^{2+}$ or greater, is not palatable for drinking water purposes, humans obtain most of the calcium and magnesium they need from food, including dairy products and meats (World Health Organization, 1993).

Many industrial applications, especially those requiring heated water, have a narrow window of allowable water hardness. These industries include power generation, printing and photofinishing, pulp and paper manufacture, and food and beverage processing.

## Impacts

Very hard waters, >180 mg/l $CaCO_3$, especially when heated,

deposit scale (calcium and magnesium carbonates) in pipes, boilers, and distribution systems, which decreases the life of the equipment. Calcium and magnesium ions also form insoluble compounds with soap. These compounds or complexes appear as soap scum, flakes, or curd (see equation below).

**Common soap scum (precipitate):**

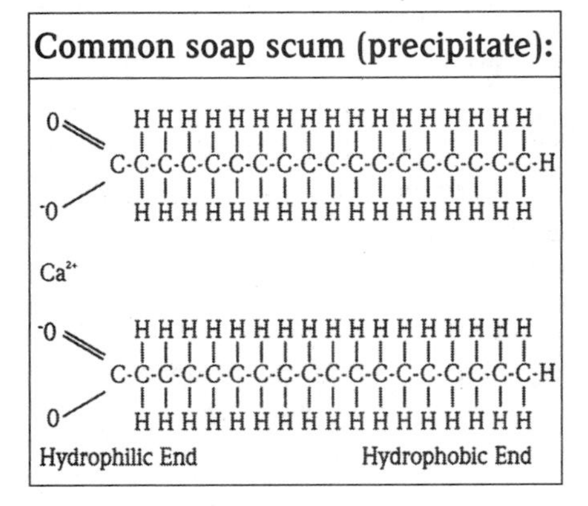


The soap scum that forms with hard water reduces the effectiveness of cleaning in various ways. It clings to laundry, leaving it with a dull gray appearance and a stiff, scratchy feel. Soap scum causes spots by clinging to shower doors, sinks, faucets, dishes and glasses. Hair washed with hard water over a long period often appears dull and lifeless. Because it is difficult to wash away, soap scum can irritate skin. Hard waters also reduce the lather and effectiveness of soap, thus increasing the amount of detergent used in cleaning.

In contrast, very soft waters may be corrosive, or "hungry" for ions. Corrosion involves the dissolution and transport of heavy metals such as cadmium, copper, lead and zinc. Very soft waters (0–60 mg/1 $CaCO_3$) can corrode those metals of pipes, boilers and distribution systems. The degree of corrosion depends on the pH, alkalinity, and dissolved oxygen levels of the water. These metals then appear in our drinking water as well as in the environment (World Health Organization, 1993).

While there are adverse health effects associated with heavy metal consumption, drinking hard water appears to have no adverse effects on humans. Fish and aquatic life can also benefit from hard water in their environment. Heavy metals such as lead (Pb), arsenic (As), mercury (Hg), and cadmium (Cd) will form compounds (complexes) with the anions associated with hardness, such as carbonate ($CO_3^{2-}$), creating insoluble compounds.
These insoluble compounds prevent the metals from being taken up by fish and other aquatic organisms via their gills, adsorbed through body tissues, or ingested. (Murphy, 2000).

### Possible Remedies

There are several means of reducing water hardness. Corrective measures utilize substances known as antiscalants and dispersants, and technology called cation exchange.

Antiscalants resist scale formation completely or allow scale to form deposits that are easily carried away by flowing water. Adding antiscalants such as washing soda, (also know as borax, sodium carbonate, or trisodium phosphate) to dishwater historically was a common remedy for hard water problems. These substances all react with calcium and magnesium, forming calcium and magnesium complexes (either carbonates or phosphates) that do not bind to fabrics and dishes.

Later, synthetic laundry and dishwashing detergents (or surfactants) were developed to adsorb to the soil or oil particle surface and carry it away, using the dispersant technique. Dispersants allow formation of scale that bears a net negative charge, which tends to repel other scale particles and prevent scale growth. Unfortunately, many of the early detergents contained phosphates ($PO_4^{3-}$), which were later found to cause environmental problems such as excessive algae growth, or eutrophication, in lakes. Phosphates have since been removed from dispersant surfactants. They have been replaced by polyacrylate alternatives that do not contribute to eutrophication; however, they are not biodegradable.

*Water hardness can damage appliances and discolor clothes. American Water Works Association*

A recently developed technology applies compounds called polyaspartates, that mimic the antiscalant (resisting scale formation) and dispersant properties of polyacrylates. Further, polyaspartate is broken down naturally (biodegrades) in the environment. Polyaspartates are now used in detergents in Europe but not in detergents in the United States (Donlar Corporation, n.d.).

Finally, softener units were created to soften water coming into homes or businesses. Softeners apply cation exchange technology, in which incoming hard water passes through a resin with a crystal or bead-like structure loaded with sodium ions ($Na^{2+}$). As hard water passes through this resin, these sodium ions are replaced by the calcium and magnesium ions in a process known as cation exchange. Water emerging from the softener contains sodium ions instead of calcium and magnesium ions. Eventually, the resin's sodium sites are completely replaced with calcium and magnesium, and the resin must be recharged. Recharging is a simple process done automatically by a water softener unit.

*Scale buildup in a water pipe due to hardness. Courtesy: Hach Company*

www.

### Case Study

When high levels of calcium, magnesium, and other polyvalent cations are dissolved in water, the water is considered *hard*. The calcium and magnesium tend to precipitate out of the water causing scale buildup when certain parameters, such as temperature, change. This scale can coat the insides of pipes, boilers, water heaters, and condensers, eventually rendering them unusable. Hard water also reduces the effectiveness of soap by producing a sticky soap scum, which can cause dull hair, stiff clothing, and spots on glasses (see Impacts).

Water softener units are a good solution to correcting hard water in homes. However, because of the quantity and quality of water required by industry, water softeners are often impractical for industrial applications. Instead, to remove the hardness that can deposit scale and damage water system equipment, most industries use chemicals such as antiscalants and dispersants. One such chemical, polyacrylate, is commonly used for industrial applications. While this product is non-toxic, it is not biodegradable and does not evaporate. When added to water to prevent scale formation, polyacrylate flows with the water to the wastewater treatment facility. There it precipitates out of the water to form a non-biodegradable sludge that must be transported to a landfill where it remains indefinitely (Narsavage-Heald, n.d.).

Recently, industry has discovered the use of polyaspartate as a substitute for polyacrylate. Polyaspartates are naturally occurring biopolymers produced by oysters to control the growth of their shells (U.S. Congress, 1993). Polyaspartate prevents the deposition of calcium carbonate in some places, thus allowing the shell to grow in its characteristic shape.

As such, polyaspartate appears to be an ideal antiscalant: it is non-toxic, hypoallergenic, and biodegradable. It breaks down to carbon dioxide and water in the environment (Narsavage-Heald, n.d.), eliminating the need to transport and store by-product sludge in a landfill. In addition, the only by-product created in the production of polyaspartates is steam, which is re-used for further production. For these reasons, the creators of polyaspartates received "the first Green Chemistry Challenge Award from the Environmental Protection Agency in 1996" (Ashley, 2002). This product has only recently been introduced in the United States but is widely used in European and other foreign countries.

This new product application illustrates that there are environmentally friendly solutions to hard water problems for both domestic and industrial water uses.

### Test Kit Activities

The day before the experiment, prepare hard water by adding limestone, concrete, or antacid tablets to tap water. Place

four funnels cut from plastic 2-liter bottles in a rack made from an upside-down cardboard box with holes cut in the bottom. Position four jars (which will later need lids) under the 2-liter bottle funnels. Place a coffee filter in each funnel. Fill one funnel one-third of the way with sand. Fill the second funnel one-third of the way with a bath water softener containing sodium hexametaphosphate (e.g., Calgon®). Fill the third funnel one-third of the way with a cation-exchange resin (Optional–can be ordered from www.hach.com, Hach item number: 26660-00, Resin Regeneration Kit, approx: $26.50). Finally, leave the last funnel empty, except for the filter, as a control.

Pour 5 ml of hard water through each funnel and collect it in the jars. Measure the hardness of the water in each jar and record the results in a data table. Add one drop of liquid Ivory hand soap (not liquid detergent) to each jar. Stir each water sample gently, cleaning the stirring device between each sample.

Compare the cloudiness or turbidity of each sample, and record your results in a data table. The more turbid the water, the more the soap dispersed. Soap that is highly dispersed cleans better. Next, place a lid on each jar and, one at a time, shake vigorously and measure the height of the lather. Water that has been softened will lather more than water that is hard. Activity adapted from *ChemCom: Chemistry in the Community* (American Chemical Society, 2002, p. 78).

### References

American Chemical Society. 2002. *ChemCom: Chemistry in the Community.* New York, NY: W.H. Freeman and Company.

Ashley, S. 2002, April. It's Not Easy Being Green. *Scientific American.* Retrieved on April 29, 2002, from the website: http://www.sciam.com/techbiz/0402innovations.html

Briggs, J., and J. Ficke. 1977. *Quality of Rivers of the United States.* Retrieved on November 29, 2001 from the Unites States Geological Survey web site: http://water.usgs.gov/owq/Explanation.html

Brunlow, A. 1996. *Geochemistry, 2nd Edition.* Upper Saddle, NJ: Prentice Hall.

Donlar Corporation. n.d. *Company Profile.* Retrieved on April 29, 2002, from the website: http://donlar.com/cp-profile.cfm

Hach Company. 1997. *Water Analysis Handbook, 3rd Edition.* Loveland, CO: Hach Company.

Murphy, S. 2001. *Interpretation of Boulder Creek Watershed Total Dissolved Solids Data.* Retrieved on December 3, 2001, from the BASIN web site: http://bcn.boulder.co.us/basin/data/NUTRIENTS/info/Hard.html

Narsavage-Heald, D. n.d. *Thermal Polyaspartate as a Biodegradable Alternative to Polyacrylate and Other Currently Used Water Soluble Polymers.* Retrieved on April 29, 2002, from the website: http://academic.uofs.edu/faculty/CANNM1/polymer polymermodule.html

Oram, B. n.d. *Water Hardness.* Retrieved on November 29, 2001 from the Wilkes University Center for Environmental Quality GeoEnvironmental Science and Engineering Department web site: http://wilkes1.wilkes.edu/~eqc/hard1.htm

U.S. Congress, Office of Technology Assessment. 1993. Biopolymers: Making Materials Nature's Way, OTA-BP-E-102. Washington DC: U.S. Government Printing Office. Retrieved on April 30, 2002, from the website: http://www.ota.nap.edu/pdf/data/1993/9313.PDF

Water Quality Association. 2000. *Water Hardness Classifications.* Retrieved on January 7, 2002 from the Water Quality Association web site: http://www.wqa.org/sitelogic.cfm?ID=362

World Health Organization. 1993. *Guidelines for Drinking-Water Quality, 2nd Ed. Vol. 1. Recommendations.* Retrieved on January 3, 2002, from http://www.who.int/water_sanitation_health/GDWQ/Chemicals/hardness.htm

# Nitrate

**Forms**

$N_2$, $NH_3$, $NH_4^+$, $N_2O$, $NO_2^-$, $NO_3^-$

**Sources**

Natural
- 79% of air
- lightning
- nitrogen-fixing bacteria
- blue-green algae

Human
- sewage
- nitrogen-fixing crops
- fertilizers and animal wastes
- burning fossil fuels

**Relevance**

Human/Animals/Plants
- metabolism, growth

**Impacts**

Human health
- Increased crop quality, yield
- Blue baby syndrome (methemoglobinemia)

Environmental health
- Less area needed for crops
- Eutrophication of ponds and lakes

**Possible Remedies**
- Efficient fertilizer techniques
- Wetland restoration

### Primary Importance

Nitrogen is found in the cells of all living things and is a major component of proteins. Nitrogen is unavailable for plant use in its most common form, atmospheric nitrogen ($N_2$). Therefore, the bioavailable species, nitrate, often becomes a limiting nutrient for plant growth.

### Technical Overview And History

Nitrogen was discovered by Daniel Rutherford in 1772, in Edinburgh, Scotland. He isolated nitrogen from air by removing oxygen and carbon dioxide. He showed that nitrogen alone could not support life or combustion. This led French chemist, Antoine Laurent Lavoisier to name the gas *azote,* meaning *without life.* French chemists still use the name azote. Nitrogen comes from the Greek words *nitron,* meaning *natural soda,* and *gennaoo,* meaning *to produce.*

For industrial uses, nitrogen is extracted from air. First air is passed through a caustic soda to remove carbon dioxide. Then the nitrogen and oxygen gas mixture is passed over heated copper or iron turnings to remove the oxygen. Nitrogen gas can also be isolated from ammonium nitrite upon heating.

Nitrogen is essential for life; it is found in the cells of all living things (Arms, 1979). Nitrogen exists in several forms, including gaseous or molecular nitrogen ($N_2$), ammonia ($NH_3$), ammonium ($NH_4^+$), nitrous oxide ($N_2O$), nitrite ($NO_2^-$), and nitrate ($NO_3^-$).

Molecular nitrogen is plentiful, making up approximately 79% of the air we breathe. However, it is not a usable form for most plants and animals because of the very strong covalent bond between the two nitrogen atoms. Fortunately, there are several ways to "fix" molecular nitrogen, or bond it to hydrogen or oxygen, to make it readily useable. Once nitrogen is fixed, it is quickly consumed by the plant community, which, in turn, feeds the animal community.

### Sources

The nitrogen cycle refers to the movement of, and associated changes in, forms of nitrogen as it cycles through an ecosystem. Algae and bacteria aid in nitrogen fixation in two ways. First, blue-green algae *(Azotobacter, Beijerinickia)*, the most common form of algae, absorb molecular nitrogen and convert it to ammonia. Second, gaseous nitrogen is fixed in the symbiotic relationship between bacteria and legumes (*Rhizobia*) such as clover, peas, beans, and alfalfa. These legumes have specialized nodules on their roots that harbor nitrogen-fixing bacteria.

When plants and animals die and decay, certain bacteria and fungi in the soil break down the large protein molecules of this organic matter to release ammonium and nitrates, while fueling their

own metabolism from the oxidation. Excretions of aquatic organisms also contribute ammonia to water. Even the heat from lightning is able to break the covalent bond of molecular nitrogen and convert it to bioavailable forms. Dissolved in water, ammonia and ammonium are "freely interchangeable" (Arms, 1979, p.669); that is, an equilibrium exists between the two species that is temperature and pH dependent.

Nitrifying bacteria (*Nitrosomonas, Nitrobacter*) then convert ammonium to nitrites, which are short-lived in the environment and are further converted to nitrates, all in the process called nitrification. While nitrification removes ammonia and nitrites that, in large quantities, are toxic to plants and animals it also has an unfortunate consequence.

Since ammonium is positively charged, it is "readily absorbed onto the negatively charged clay colloids and soil organic matter, preventing it from being washed out of the soil by rainfall. In contrast, the negatively charged nitrate ion is not held on soil particles and so can be washed down the soil profile in a process termed leaching. Valuable nitrogen lost in this way reduces soil fertility. The nitrates can then accumulate in ground water, and ultimately drinking water" (Deacon, 2001, p. 11).

Eventually some of the leached nitrates return to the atmosphere by denitrification. This process, which reduces nitrates to intermediate nitrites and eventually nitrogen or nitrous oxide, also occurs in anaerobic soils with the help of microorganisms (*Pseudomonas, Alkaligenes, Bacillus*).

If nitrates do not undergo denitrification, they are likely utilized by plants for cell growth. Animals then eat the plants (or eat animals that have eaten plants) to obtain the nitrogen they need for metabolism, growth, and reproduction. The cycle continues with animal excrement, death and decay.

There is an ever-growing supply of nitrates from human (anthropogenic) sources being interjected into this nitrogen cycle. The largest human contribution to the system is industrially fixed nitrogen used in fertilizers (Vitousek et al., 1997). Fertilizers are easily carried via water, either from irrigation water leaching

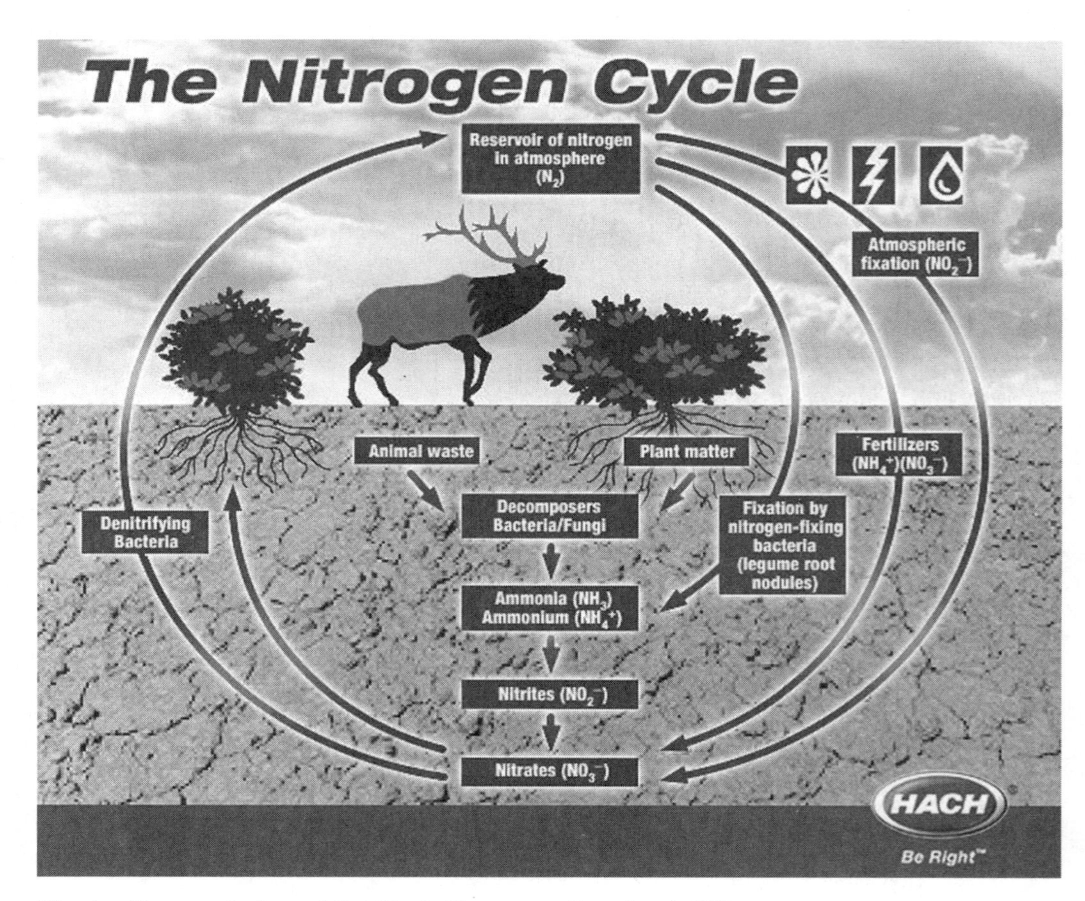


*Used with permission of the Hach Company, Loveland, CO.*

into the ground water or from runoff entering the natural system.

Some geologic formations hold long-term stores of usable nitrogen now being mined for coal and oil. If not for human intervention, this nitrogen would remain in storage indefinitely, but the burning of fossil fuels releases it. In addition, inadequately treated wastewaters from improperly functioning septic systems and illegal sewer connections contribute nitrogen to the environment. Animal wastes from dairies, feedlots, and barnyards; the draining of wetlands; clearing of land for crops; and the burning of forests, wood fuels and grasslands all are human interventions into the natural nitrogen cycle.

### Relevance

Plants and animals use ammonium and nitrate as building blocks for proteins, amino acids, and nucleic acids (Pidwirny, 2000). The bio-available forms of nitrogen are often called "nutrients." They, along with phosphates and potassium, are the predominant nutrients needed for healthy plant growth. However, because nitrates are not the most abundant form of nitrogen, they may be a limiting nutrient or resource in plant growth. Plants limited by phosphate availability still need nitrates and potassium for healthy growth. Animals then eat the plants, or eat animals that have eaten plants, to obtain the nitrogen they need for metabolism, growth, and reproduction

### Impacts

The increase in nitrates due to human action has caused an increase in health-related disorders in human and aquatic organisms. When humans consume nitrates leached into drinking water sources, a serious condition called methemoglobinemia can develop. Infants, pregnant women, and elderly people are much more susceptible to this condition, which is a result of the transformation from nitrate to nitrite in the digestive system. This nitrite oxidizes the iron in the hemoglobin of red blood cells to form methemoglobin (Oram, p.2). Methemoglobin is unable to carry oxygen like hemoglobin; therefore cells do not receive their required oxygen. This lack of oxygen causes veins and skin to appear blue, hence the term "blue baby syndrome." To prevent these illnesses, the USEPA has established a maximum contamination level (MCL) for nitrate of 10 mg/l $NO_3$-N. (Fish exhibit a similar reaction to nitrites, referred to as "brown blood disease.")

The environment has also been impacted by an increase in nitrates. Since nitrogen is a limiting nutrient for plants, an increase in usable nitrogen can cause increased plant and algae growth. This effect is known as eutrophication. When there are more aquatic plants and algae near the surface, plants that live deeper in the water column are starved of light, die, and are decomposed by oxygen-depleting bacteria. The impact continues when other oxygen-dependent organisms in the system suffocate.

The acidification of lakes has been attributed in part to the increase in nitrates in the environment. Nitrogen oxides, components of some industrial emissions, contribute to acid deposition in the form of rain, snow, fog, and mist. In addition, some soils that once buffered acid rain before it entered our waterways have become saturated with nitrates from human actions and have lost this buffering capacity.

The application of nitrates in fertilizers has also been beneficial. Nitrates in fertilizers have increased agricultural production to meet the food demands of a growing population throughout the world. Using fertilizers to increase crop yields on existing agricultural lands preserves other land for forests and animal habitats. According to the International Fertilizer Industry Association, "fertilizers account for at least a third of all crop yields, sometimes up to half or more in soils depleted of plant nutrients" (n.d., p. 1). Fertilizers increase both crop quality and yield by satisfying the specific vitamin and mineral requirements for plants. Humans consuming these crops benefit from the high vitamin and mineral content. When fertilizers are properly and efficiently applied to soils, they accumulate, so less fertilizer is required in the following growing season.

### Possible Remedies

Fortunately, there are several ways that humans can reduce the contribution of nitrates to the environment.

*Proper fertilizer application can reduce nitrate levels in nearby waterways. American Water Works Association*

The best remedy is to properly follow fertilizer application instructions and apply only the amounts needed.

Wetlands and riparian areas can also help slow the fertilizers leaching into ground water. "Restoration of wetlands and riparian areas, and even construction of artificial wetlands, have shown to be effective in preventing excess nitrogen from entering waters" such as ground water, streams, estuaries, and coastal waters (Vitousek et al., 1997, p. 12).

Water treatment plants can apply certain techniques to reduce nitrate concentrations in source water. These techniques include reverse osmosis and dilution, or blending. This practice involves blending low-nitrate water, usually purchased from a different source, with the high-nitrate water to produce a water that is acceptable for consumption. However this technique does not remove nitrate from the water.

**Case study**

"The Raccoon River is the most nitrogen-polluted river in America's heartland" (Dewar et al., 2000). A state researcher indicated that two of Iowa's major rivers, the Raccoon River and the Des Moines River, have nearly tripled their nitrate loads since the 1940's (Beeman, 2001). These are the two main water sources for the Des Moines Water Works (DMWW) and the residents of Des Moines. Jerry Hatfield, representing the National Soil Tilth Laboratory at the U.S. Department of Agriculture Research Service, was quoted as saying that 90 percent of the nitrates in Iowa waterways come from agriculture-related enterprises (The Des Moines Register, 2001, July 1). Sources include primarily fertilizers, livestock manure, soybean crops, and certain soil tilling methods.

Because of health concerns, the DMWW began notifying their customers in1989 when nitrates in the water exceeded the United States Environmental Protection Agency (USEPA) Maximum Contaminant Level (MCL) of 10 mg/1 . High levels of nitrates in drinking water have been linked to several human health conditions, including blue baby syndrome, non-Hodgkin's lymphoma, and miscarriages (The Des Moines Register, 2001, July 1).

The situation worsened when nitrate concentrations exceeded the MCL in the Des Moines River more than 130 days in 1991 and 1992. Likewise, nitrate concentrations exceeded 10 mg/l more than 140 days in the Raccoon River in 1991 and 1992 (Raccoon River Watershed Project, 2000).

This situation led to a plan to build a nitrogen removal facility. In 1992, after three years of planning and spending $3.7 million, the DMWW nitrogen removal facility began effectively removing nitrates from drinking water. This facility utilizes eight vessels—with room for two additional vessels if necessary—each about 14 feet tall, 11 feet in diameter, and weighing 11,000 pounds.

Each vessel contains both ion exchange resin and gravel as a medium to filter the nitrates. Initially, the ion exchange resin is treated with sodium chloride. Then, water containing nitrates is passed through the resin and nitrate ions are exchanged for chloride ions. The nitrate ions remain in the resin, while the emerging water contains chloride ions. Once the resin exchange sites are full of nitrogen ions, sodium chloride is flushed through the system. This flushing produces nitrate-laden water that is then diluted and discharged into the Raccoon River. Such an operation can cost up to $4,500 per day (Peckumn, 2001).

Since this is a very expensive facility to operate and the nitrates are not permanently removed from the watershed, the DMWW also is educating the public about ways to protect its waters. Its program includes formal in-school education, a monthly newsletter mailed with residents' water bills, a quarterly teachers' newsletter, and a web page with tips for proper fertilizer use, other conservation measures, and general environmental stewardship (Peckumn, 2001).

### Test Kit Activities

Place 5 clean funnels in a rack, and line each with filter paper (plastic 2-liter bottles can be modified to serve as funnels, with an upside down cardboard box with holes cut in it to serve as a rack). Place a small container under each funnel to capture the sample water. Fill each funnel with soil (not sod) to a depth of one-third. Add nothing to the control sample. Add the same amount of fertilizer to each of the additional funnels as follows: to funnel 1 apply solid commercial fertilizer (e.g, Miracle Grow®, Shultz®); to funnel 2 apply the same amount of liquid commercial fertilizer (e.g, Miracle Grow®, Shultz®); to funnel 3 apply a commercial steer manure to the surface; and to funnel 4 apply plant residue (e.g., compost, leaves, grass clippings, etc.).

Water each sample with the same known volume of water. Test the water that has percolated through the soil and into the collection container for nitrates and phosphates. Record results in a data table. Repeat the following day, and again for several days. Does one nutrient leach through soil more than the other? Graph the results to illustrate how nutrients move through soil over time.

### References

2001, July 1. Waterworks Struggle to Cope. *The Des Moines Register.* [Electronic version] http://www.aw-waneb.org/articles/2001news/waterworks.html

Arms, K. & P. Camp. 1979. *Biology.* New York, NY: Holt, Rinehart & Winston.

Beeman, Perry. 2001, March 30. State Finds High Nitrate Levels. *The Des Moines Register.* [Electronic version] http://www.water.iastate.edu/march30a.html

Deacon, J. 2001. *The Microbial World: The Nitrogen Cycle and Nitrogen Fixation.* Retrieved on October 8, 2001 from the Institute of Cell and Molecular Biology web site: http://helios.bto.ed.ac.uk/bto/microbes/nitrogen.htm

Dewar, H. & T. Horton. 2000, September 25. Nitrates are elusive suspect in assortment of illnesses. *Baltimore Sun.* [Electronic version] http://www.baltimoresun.com/news/nationworld/bal-nitrogen-sidep2.story

Havlin, J., J. Beaton, S. Tisdale, and W. Nelson. 1999. *Soil Fertility and Fertilizers: An Introduction to Nutrient Management, 6th Edition.* Upper Saddle River, NJ: Prentice Hall.

International Fertilizer Industry Association. n.d. *Fertilizer Use and the Environment.* Retrieved on March 25, 2002, from the website: http://www.fertilizer.org/ifa/ab_act_position4.asp

Mitchell, M. & W. Stapp. 1997. *Field Manual for Water Quality Monitoring: An Environmental Education Program for Schools.* Dubuque, IA: Kendall/Hunt Publishing Co.

Oram, n.d. *Nitrates and Nitrites in Drinking Water.* Retrieved on October 2, 2001 from the Wilkes University Center for Environmental Quality GeoEnvironmental Science and Engineering Department web site: http://wilkes1.wilkes.edu/~eqc/nitrate1.htm

Peckumn, G. Education Specialist, Des Moines Water Works. Correspondence on October 18, 2001.

Pidwirny, M. 2000. *The Nitrogen Cycle.* Retrieved on October 8, 2001 from the Okanagan University College, Department of Geography web site: http://www.geog.ouc.bc.ca/physgeog/contents/9s.html

United States Environmental Protection Agency. 1997. *Volunteer Stream Monitoring: A Methods Manual.* United States Environmental Protection Agency.

Vitousek, P.M., J. Aber, R. Howarth, G. Likens, P. Matson, D. Schindler, W. Schlesinger, and G.D. Tilman. 1997. *Human Alterations of the Global Nitrogen Cycle: Causes & Consequences.* Retrieved on October 8, 2001 from the Ecological Society of America web site: http://esa.sdsc.edu/tilman.htm

# pH

**Primary Importance**

pH affects the solubility, and thus bioavailability, of many substances in water, some of which can be toxic to humans and aquatic life.

**Technical Overview And History**

A water molecule is made up of a hydrogen ion ($H^+$) and a hydroxide ion ($OH^-$): $H^+ + OH^- = H_2O$.
The hydrogen ion concentration determines the pH of a solution. The term pH comes from the French "pouvoir hydrogène," literally, "hydrogen power." pH is generally referred to as hydrogen ion concentration or activity.

An acid is a solution with more hydrogen ions than hydroxide ions. When an acid is dissolved in water, the net effect is an increase in the water's hydrogen ion concentration. A base, with more hydroxide ions than hydrogen ions, increases the concentration of hydroxide ions when it is dissolved in water. A neutral solution has an equal number of hydrogen and hydroxide ions.

The pH test measures the hydrogen ion concentration and allows us to infer how acidic or basic a substance is. Since hydrogen ion and hydroxide ion concentrations are very small absolute numbers, scientists developed the pH scale to make reporting and interpreting the numbers easier. Here's how the pH scale is derived:

**Sources**

Natural
- Carbon dioxide
- Limestone rocks

Human
- Acid precipitation
- Acid mine drainage
- Excess nutrients

**Relevance**

Human
- Affects metals solubility in drinking water pipes
- Range drinking water is safe (pH 6.5 to 8.5)

Animals/Plants
- Affects solubility of metals and nutrients
- Range within which most animals can survive (pH 5 to 9)

**Impacts**

Human health
- Beyond normal range unsafe to drink

Environmental health
- Ammonia toxicity in animals at high pH
- At low pH, toxic metals (e.g., aluminum) become soluble
- Beyond normal range, fatally disrupts osmotic balance

**Possible Remedies**
- Reduce NO and $SO_2$ emissions
- Treat mine acid drainage

The product of the hydrogen ion concentration and hydroxide ion concentration is equal to $1.0 \times 10^{-14}$, at 25°C.
$[H^+][OH^-] = 1.0 \times 10^{-14}$
This value is known as the ion product constant for water, a factor that varies only slightly with temperature.

Because the product of $[H^+]$ and $[OH^-]$ is a constant, if one value increases, the other must decrease. For example, if an acidic substance increases the concentration of hydrogen ions in solution, the concentration of hydroxide ions must correspondingly decrease. In a neutral solution, such as pure water, the concentrations of hydrogen ions and hydroxide ions are equal:
Hydrogen ion concentration, $[H^+] = 1.0 \times 10^{-7}$ moles/liter Hydroxide ion concentration, $[OH^-] = 1.0 \times 10^{-7}$ moles/liter (Note: the brackets represent *concentration*, and a mole is a counting unit in chemistry equal to $6.02 \times 10^{23}$, the number of atoms in one liter of water)

pH is defined as the negative of the logarithm of the hydrogen ion concentration. Therefore, the concentration of hydrogen ions in a neutral solution, in terms of the pH scale, is:
$pH = -\log[H^+] = -\log(1.0 \times 10^{-7}) = -(-7) = 7$. It is much easier to report the pH of neutral water as 7, rather than $1.0 \times 10^{-7}$ moles/liter. Note: because of its logarithmic nature, pH cannot be averaged in the same way as other data.

Here is how the pH scale works:
The pH scale ranges from 0 to 14.
An acidic solution has a pH less than

7, while a basic solution has a pH greater than 7. Note that an acidic solution has a lower pH–less than 7, but a greater hydrogen ion concentration. This inverse relationship is reflected by the negative sign in front of the log in the pH definition equation.

Since this is a logarithmic scale, for every one unit of change in pH there is a ten-fold change in the hydrogen or hydroxide ion concentration of the solution. For example, rain is slightly acidic with a pH of about 6, while acid rain is ten times more acidic with a pH of about 5. Many soft drinks are 1000 times more acidic than rain, with a pH of approximately 3. Likewise, a solution with a pH of 10 is one hundred times more basic than a solution with a pH of 8.

### Sources

The carbonate system is one of the most prominent equilibrium systems in natural waters. Carbon dioxide ($CO_2$), carbonic acid ($H_2CO_3$), bicarbonate ($HCO_3^-$), and carbonate ($CO_3^{2-}$) make up this carbonate system. The presence or absence of these carbonates can contribute to pH changes in natural waters.

Aquatic plants manipulate the carbonate system and, therefore pH, via photosynthesis and respiration. Photosynthesis uses carbon dioxide that is dissolved in water as carbonic acid ($CO_2 + H_2O = H_2CO_3$). During plant growth, carbonic acid is removed from water and the pH increases. As aquatic plants respire and bacteria consume decomposing matter, they release carbon dioxide that reacts with water to form carbonic acid. Therefore, at night while plants are respiring, the pH decreases with the increase in carbonic acid. These diurnal changes in pH are measurable but are not significant enough to affect the health of aquatic life.

The surrounding geology also directly impacts the carbonate system and natural water pH. Weathering and erosion of limestone and other carbonate rocks by slightly acidic natural water quickly dissolves these rocks, releasing carbonates. These carbonates act as natural buffers against large pH fluctuations (see Alkalinity).

The landscape in a watershed can influence the pH of its waters. Watersheds that contain bogs, marshes, or pine forests tend to support waters with a lower pH. Sphagnum moss and pine needles are slightly acidic. Decaying vegetation produces organic acids that leach into the ground or join runoff and flow into nearby waters, thus reducing the pH.

Burning of fossil fuels and other human activities eventually contributes to changes in pH. Coal-fired power plants and automobiles emit

*Abandoned mine shaft with mine acid drainage flowing out of it. Courtesy: Donald Bain; University of California Berkeley.*

nitrogen oxides ($NO_2$, $NO_3$) and sulfur dioxide ($SO_2$) into the air. These compounds react with water vapor in the air to form nitric acid ($HNO_3$) and sulfuric acid ($H_2SO_4$): $2\ SO_2 + O_2 = 2\ SO_3$ and then $SO_3 + H_2O = H_2SO_4$. These acids combine with moisture in the atmosphere and fall as acid precipitation–either acid rain or acid snow. Acid snow accumulates over the winter, and releases a potentially toxic dose of acidic runoff in the spring. Without the natural buffering contributed by the surrounding geology, the pH of a water system can decrease to fatal levels. This phenomenon has occurred in the northeastern United States, Finland, and Sweden (Mitchell, 1997).

Mining, chemical spills, thermal pollution, urban runoff, sewage effluent, and agricultural runoff also can affect the pH of water systems. Mining for coal, gold, silver, and other metals often expose sulfide minerals to the atmosphere where they react with water to form sulfuric acid. This acid mine drainage combines with surface and ground water and lowers pH.

### Relevance

In general, most aquatic organisms survive in waters between a pH of 5 to 9. Beyond this range, the diversity of species decreases, as pH fluctuations and the associated solubility and bioavailability changes will stress or even kill aquatic life.

Water with low pH will corrode lead, cadmium, copper, or zinc pipe and solder in drinking water. Further, a change in pH can cause a chain reaction of events. Thermal pollution, agricultural runoff, and excessive nutrients result in decreased pH, more soluble (and available) phosphorus, further plant and algae growth, and ultimately reduced dissolved oxygen.

### Impacts

Most aquatic organisms have adapted to a specific pH range beyond which they simply cannot survive. Waters with a pH less than about 5 are too acidic for human consumption and are unable to support most aquatic life. Waters that are too basic also threaten humans, plants, and animal populations. Waters with a pH greater than 9 can dissolve organic materials, including animal scales and skin (American Chemical Society, 2002). For this reason, the Environmental Protection Agency has established a secondary (not legally enforceable) drinking water standard pH range of 6.5 to 8.5. Typically, the pH of natural water in the U.S. is in this range, although there are exceptions.

In the pH range of natural waters, ammonia reacts with water to form ammonium and hydroxide ($NH_3 + H_2O = NH_4^+ + OH^-$) both of which are benign to aquatic organisms. In waters with pH greater than 9, however, ammonia does not react with water. The resulting unreacted ammonia is harmful to many aquatic organisms.

Ultimately, increasing or decreasing pH beyond the natural range can

**The pH of Common Items**

| Substance | pH or pH range |
|---|---|
| Hydrochloric Acid | 0.0 |
| Stomach Acid | 1.0–3.0 |
| Lemon Juice | 2.2–2.4 |
| Vinegar | 2.4–3.4 |
| Cola | 2.6 |
| Grapefruit | 3.0–3.2 |
| Acid Rain | 4.0–5.5 |
| Natural Rain | 5.6–6.2 |
| Milk | 6.3–6.7 |
| Pure Deionized Water | 7.0 |
| Sea Water | 7.0–8.3 |
| Baking Soda | 8.4 |
| Milk of Magnesia | 10.5 |
| Household Ammonia | 11.9 |
| Sodium Hydroxide | 13.0–14.0 |

*Table compiled from Ebbing, 1990, and Tillery, 1996*

result in a species shift as organisms adjust to the changes, migrate to healthier waters, or die. Such a pH alteration can have severe repercussions on recreation, the food chain, and the long-term survival of species.

### Possible Remedies

As we learn more about human impacts on natural systems, we are better able to adjust our actions. For example, the Clean Air Act Amendment of 1990 aims to reduce acid precipitation by limiting the emissions of sulfur dioxide and nitrogen oxides (Lynch et. al, 1997).

Additionally, various remediation technologies in place today reduce acid mine drainage from both abandoned and functioning mining operations. These efforts include manufactured wetlands with limestone treatment and proper disposal of the contaminated soils or substrates. Monitoring industrial effluent helps reduce pollution and pH changes.

### Case Study

Acid deposition–acid rain, snow, sleet, and hail–has significantly impacted soil chemistry, forest vegetation, water quality, and aquatic organisms in the northeastern United States. In fact, the harmful effects of acid rain are greater than predicted (Driscoll, et. al, 2001).

The Clean Air Act Amendment (CAAA) of 1990 was implemented because of the detrimental effects of acid precipitation on watersheds. The impacts were particularly significant in the Northeast because of the dominant geology and landscape and the relative volume of sulfur and nitrogen oxide emissions.

Acid precipitation occurs where sulfur dioxide and nitrogen oxides react with oxygen, and then moisture, in the atmosphere to produce sulfuric and nitric acids. These acids are not as significant of a threat if carbonate rocks occur in the area. Carbonates dissolve in water and then reduce or buffer the effects of acids. In this environment, acid rain will not decrease the pH of the receiving waters.

Igneous rocks, such as granite, which is primarily composed of the minerals quartz, feldspar, and mica is the dominant geology of the Northeast. Instead of dissolving in water, these minerals break into finer and finer pieces that eventually make up sand and clay. They do not contribute carbonates that buffer water from acid precipitation (see Alkalinity).

To document the impacts of the CAAA, scientists have studied the interactions between geology and acid precipitation in the Northeast over time. Various independent researchers and national agencies including the Environmental Protection Agency (USEPA), the National Oceanic and Atmospheric Administration (NOAA), and the National Atmospheric Deposition Program (NADP) have monitored acid precipitation since the 1970s.

One such organization, the Hubbard Brook Research Foundation (HBRF), has concluded that, while sulfur emissions have decreased since the passage of the CAAA, nitrogen emissions continue to increase. Additionally, the long-term effects of acid deposition have reduced the ability of

*Burning of fossil fuels and other human activities can eventually contribute to changes in pH through acid precipitation. Courtesy: United States Environmental Protection Agency*

the ecosystem to recover from such impacts, despite reduced emissions.

HBRF researchers reported several ongoing impacts to soil quality throughout the Northeast. Long-term acid deposition has reduced the ability of soils to neutralize or buffer additional acid deposition, compounding the impacts to the watershed. The dissolved inorganic aluminum concentration has increased, while the concentration of essential nutrients for plant growth has decreased. Therefore, the overall growing conditions have significantly diminished, as evidenced by reduced tree growth, poor crown condition of trees, and the increase in tree mortality.

Water quality has also been affected by acid deposition. Specifically, scientists have noted a decrease in pH, a reduction in the ability of waters to buffer acid inputs (acid neutralizing capacity), and an increase in aluminum concentrations. Ultimately, these water quality factors have reduced the diversity and abundance of aquatic life (Driscoll, et. al, 2001).

Further reductions in nitrogen emissions will reduce acid deposition. As acid deposition decreases, the ecosystem will recover. Initially, water pH and the acid neutralizing capacity is expected to increase. Soil will slowly rid itself of the sulfate, nitrate and aluminum stores accumulated over decades of acid deposition. Only then will soils return to natural cation (positively charged iron) levels, and support healthy vegetative growth.

After the chemical balance of soils and water is restored, biological communities are likely to recover. Scientists anticipate macroinvertebrates will return over the course of about three years. Zooplankton would soon follow and reach natural levels in approximately ten years. Once these food sources are established, fish would return. It could take up to 20 years' to support a healthy fish population; however, fish stocking can expedite this process, assuming food sources are reliable. Finally, after about 20 years, scientists estimate trees would recover from the impacts of acid precipitation (Driscoll, et. al, 2001).

**Test Kit Activities**

In a small glass jar with a lid, collect a 10 g soil sample from directly under a pine tree. In a second small glass jar with a lid, collect an additional 10 g soil sample from at least 50 feet away from pine trees. Add 10 ml of deionized water to each of the samples. Shake the soil samples for one minute every ten minutes, and repeat three times (for a total of 30 minutes). Test the pH of each soil sample. Which sample is higher? Why? How might this influence other plants growing nearby?

**References**

American Chemical Society. 2002. *ChemCom: Chemistry in the Community.* New York, NY: W.H. Freeman and Company

Bartenhagen, K, M. Turner, and D. Osmond, 1995. *pH.* Retrieved on October 29, 2001 from the North Carolina State University WATERSHEDDS: Water, Soil and Hydro Environmental Decision Support System web site: http://h2osparc.wq.ncsu.edu

Driscoll, C., G. Lawrence, A. Bulger, Butler, T., C. Cronan, C. Eagar, K. Lambert, G. Likens, J. Stoddard, and K. Weathers. 2001. *Acid Rain Revisited: Advances in scientific understanding since the passage of the 1970 and 1990 Clean Air Act Amendments.* Retrieved on March 20, 2002, from the Hubbard Brook Research Foundation website: http://www.hbrook.sr.unh.edu/hbfound/hbfound.htm

Ebbing, D. 1990. *General Chemistry, third Edition.* Boston, MA: Houghton Mifflin Co.

Lynch, T., V. Bowersox, and J. Grimm. 1997. *Trends in Precipitation Chemistry in the United States, 1983 – 1994.* Retrieved on March 18, 2002, from the United States Geological Survey website: http://water.usgs.gov/pubs/acidrain/

Mitchell, M. and W. Stapp. 1997. *Field Manual for Water Quality Monitoring.* Dubuque, IA: Kendall/Hunt Publishing Company.

# Phosphate

### Primary Importance

"Phosphorus is an essential nutrient for all forms of terrestrial life and is one of the 17 chemical elements known to be required for plant growth" (Waskom, 1994, p. 1). In excess it causes cultural eutrophication of lakes

### Technical Overview and History

Phosphorus (P) comes from the Greek words *phôs*, meaning light, and *pharos*, meaning bearer.

Elemental phosphorus was discovered in 1669 by Henning Brand, a German merchant and alchemist.

Brand first distilled phosphorus from urine and later found it in bones. He and other researchers reacted bone material with nitric acid and with sulfuric acid to produce phosphoric acid. They then heated phosphoric acid with coal to extract elemental phosphorus (P).

This method of commercial phosphorus production continued until the end of the 19th century when James Readman developed the first electric furnace production method. Today's production of elementary phosphorus applies this newer method, using minerals instead of bones as source material (Physics Department, University of Coimbra, Portugal, n.d.).

**Forms**

$PO_4^{3-}$, $HPO_4^{2-}$, P, $H_2PO_4^-$

**Sources of Phosphates**

Natural
- Apatite (mineral)
- Elemental Phosphorus

Human
- Animal and human waste
- Industrial discharge
- Fertilizers
- Soil erosion

**Relevance**

Human
- Production of fireworks, matches, steel
- Maintain teeth
- Baking powder
- Soft drinks
- ATP, DNA, RNA

Animals and plants
- Maintain bones, shells,
- Energy production, storage, and transmission
- ATP, DNA, RNA

**Impacts**

Human Health
- Disrupts wastewater treatment

Environmental Health
- Cultural eutrophication
- Decreases dissolved oxygen levels
- Shift from aerobic to anaerobic bacteria
- Shift to pollution tolerant species
- Algal blooms/mats

### Sources

Phosphates are found in the mineral apatite, which occurs naturally in igneous, metamorphic and sedimentary rocks. Rocks with a high concentration of apatite are often referred to as phosphate rocks. In the U.S.A., there are major phosphate rock deposits in Florida, Idaho, Montana, and Tennessee. Outside the U.S.A., major deposits exist in Russia, Morocco, South Africa, and China. The natural weathering, (erosion) of these deposits produces orthophosphates ($H_2PO_4^{1-}$ and $HPO_4^{2-}$).

However, more than half of the phosphates found in lakes, streams, and rivers are the result of human activity (Hartman et al., 2000, p. 36). Discharge of animal and human wastes, industrial and domestic wastes, and fertilizers introduces phosphates into natural waters, as does increased soil erosion from the removal of vegetation. As organic phosphates break down, one of the by-products can be orthophosphate, which becomes available for use by both plants and animals.

### Relevance

Animals use both orthophosphates and organic phosphates for the metabolism of fats, carbohydrates, and proteins. Plants require phosphorus for growth and use only orthophosphates as a source for this nutrient.

However, orthophosphates and organic phosphates are produced naturally in small amounts. Consequently,

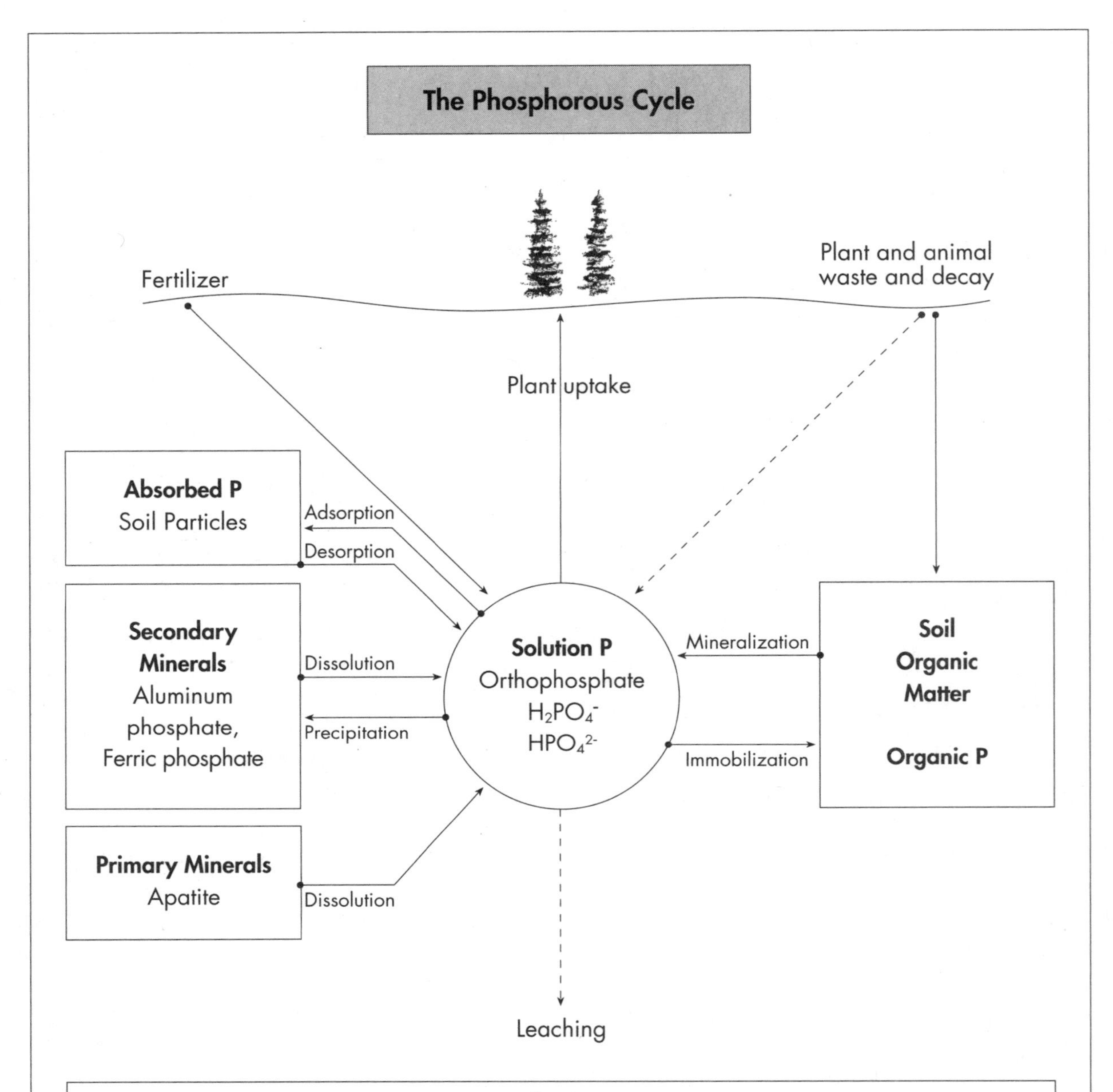


**Definitions**:

**Adsorption:** binding of P to soil particles or clay minerals in the soil.
**Desorption:** opposite of adsorption; P is released from the surface of soil particles to be used by plants or animals once again.
**Mineralization:** organic P decomposes to the bioavailable orthophosphate.
**Immobilization:** orthophosphate becomes bound in microorganisms' cells.
**Precipitation:** P from solution that becomes part of a mineral.
**Dissolution:** P that was part of a mineral that goes into solution.
**Leaching:** while infrequent, some P can leach through the soil and into the ground water.

phosphate (along with nitrate or nitrogen) is considered a limiting nutrient for plant growth. Once the limited supply of available nutrients is consumed, plants can no longer grow until that supply is replenished.

Positively charged particles (cations) in the soil, such as iron, calcium, magnesium and aluminum, have a high affinity for negatively charged orthophosphates. Combined with these cations, orthophosphates remain in the soil instead of leaching into ground water and are readily usable (bioavailable) to both plants and animals for growth. In fact, orthophosphate excess that causes increased algal growth of lakes and reservoirs is usually an indicator of human-caused pollution.

Plants and animals use phosphorus for metabolic energy production, storage, and transmission. In order to form adenosine triphosphate (ATP), a phosphorylated compound considered an energy intermediate in every living organism, there must be sufficient phosphorus available (Arms, 1979, p. 76).

Molecules containing phosphorus also make up the long chains that store and transmit genetic information, DNA (deoxyribonucleic acid) and RNA (ribonucleic acid) (Knochel, ed. 2001). Further, phosphorus acts as one of the body's natural buffers, helping maintain constant pH. Finally, an adequate supply of phosphorus is needed to produce and maintain bones, shells, and teeth.

Phosphorus and its many compounds are significant in industry. Metaphosphates (polyphosphates) are synthetic, and are used in boiler waters to prevent scale build up. Polyphosphates are also occasionally added to drinking water for corrosion control. When discharged into natural waters, polyphosphates are easily converted to orthophosphates, and are then available for plant and animal use.

Elemental phosphorus is used in the production of fireworks, matches, insecticides, steels and phosphorous bronze and phosphorus brass.

Phosphates are used in the production of dentrifices, reagents, special glasses (sodium lamps), toxic nerve gas (Sarin), fertilizers, detergents, pharmaceuticals, water softeners, fine chinaware, and baking powder.

Applications of phosphoric acid include the cleaning of metals, refining sugar, rust proofing, pickling metals, and as an additive to soft drinks.

Phosphorus is not harmful to our health, except that it may disrupt the normal coagulation processes of wastewater treatment plants and allow the release of untreated organic particles that carry harmful microorganisms (Bartenhagen et al., 1995).

### Impacts

According to the American Heritage Dictionary, eutrophic means, "having waters rich in mineral and organic nutrients that promote a proliferation of plant life, especially algae, which reduces the dissolved oxygen content and often causes the extinction of other organisms." (www.dictionary.com). One of the major impacts of unnatural amounts of phosphates and nitrates in our system is their eventual contamination of our waters. The direct result of nutrient overload is eutrophication, or excessive aquatic plant and algal growth, often at the expense of other organisms.

Lakes and ponds can become eutrophic naturally over a very long period if fed by phosphate–rich waters from phosphate rock deposits. Forest fires and volcanoes also speed up erosion rates, and therefore, potentially, the nutrient levels of a lake or pond. Under normal conditions, most of the phosphates that enter a lake are either taken up by plants or they sink to the bottom and are soon covered by sediments.

Today, however, the widespread eutrophication of lakes is a result of cultural eutrophication, or human-induced nutrient loading. Nutrients are added to aquatic systems via fertilizers from lawns and crops, wastewater treatment plant effluent, increased sedimentation from land use changes, and animal waste.

Also, wetlands that hold long-term stores of phosphates are being overcome by development. The impact of wetland destruction is two-fold: a storage area is permanently removed from the system, and phosphates stored for a long time are released

into the environment.

When lakes are unnaturally supplied with phosphates and nitrates, aquatic plants and algae readily consume the extra nutrients and grow excessively. As a general rule, aquatic plants in marine environments are limited by the availability of nitrogen. Therefore, if nitrogen is abundant, plants will grow excessively. Similarly, aquatic plants in fresh water systems are often limited by the availability of phosphorus. Hence, if phosphorus is abundant, plants will grow excessively.

Frequent algal blooms are often the first sign of cultural eutrophication, regardless of which element is limiting. Algal blooms trigger a sequence of changes:

- First, they do not allow sunlight to penetrate to the deeper aquatic plants, which then die.

- When large amounts of organic matter begin to decay, the aerobic bacteria that consume it grow vigorously and use up the available dissolved oxygen (DO).

- Depleted DO causes more plants, and certain species of fish, to die.

- Ultimately, oxygen becomes so depleted that aerobic bacteria cannot survive, and anaerobic bacteria begin to thrive. In advanced stages of cultural eutrophication there is a characteristic "rotten egg" smell from the respiration of the anaerobic bacteria.

**Which Nutrient is Limiting?**

To determine if nitrates (nitrogen) or phosphates (phosphorus) limit a body of water, the N/P ratio is used. Aquatic plants use these elements for growth at a concentration of 16 N to 1 P (Welch, 1980). Therefore, if the N/P ratio is greater that 16/1, then aquatic plants are limited by phosphorous. For example, if the N/P ratio is 28/1 and the plants need 16/1, they will run out of available phosphorus before they run out of available nitrogen. Hence, phosphorous is limiting their growth. Likewise, if the N/P ratio is below 16/1, nitrogen limits aquatic plant growth.

- Eventually, the lake or pond will fill with decaying organic matter and become a swamp or bog devoid of the aquatic organisms that once thrived there. Birds and terrestrial animals that once were dependent on the lake suffer as well.

The impact of cultural eutrophication on humans extends beyond the foul appearance and smell of the lakes. Increased occurrence of the fish-killing toxic *Pfiesteria* algal blooms along the Atlantic Coast is considered responsible for conditions ranging from impaired memory loss and confusion to respiratory, skin, and gastrointestinal problems in humans (USEPA, 2001).

It is believed these *Pfiesteria* blooms, originally discovered in the late 1980's, might be caused by nutrient enrichment–the catalyst of cultural eutrophication (Reuther, 1998; USEPA, 2001).

### Possible Remedies

Fortunately, there are ways to reduce cultural eutrophication. Mitchell and Stapp (1997) recommend the following:

**1.** Reduce use of lawn fertilizers (particularly inorganic forms) that drain into waterways;

**2.** Encourage better farming practices: low-till farming to reduce soil erosion; soil-testing to guide fertilizer application and prevent excess fertilizer from finding its way into waterways; build storage or collecting areas around cattle feedlots to prevent the runoff of phosphorus-containing manure;

**3.** Preserve natural vegetation whenever possible, particularly near shorelines; preserving wetlands that absorb nutrients and maintain water levels; enacting strict ordinances to prevent soil erosion;

**4.** Support measures to develop phosphorus-removal technology for wastewater treatment plants and septic systems; treat storm sewer wastes if necessary; encour-age homeowners along lakes and streams to invest in community sewer systems;

**5.** Require phosphate-free detergents to reduce nutrient loading. While considered effective, such remedial measures can take several years to effectively reduce eutrophication. Simply, phosphates stored near the surface of the sediments will continue to dissipate until depleted. For example, benthic (bottom-dwelling) invertebrates will disturb the bottom sediments, re-introducing phosphates into the water column. Seasonal changes to the chemistry of the water can re-dissolve phosphates that remain exposed in the sediments. Ultimately, however, the lake can recover from cultural eutrophication if nutrient loading is stopped.

### Case Study

The extreme cultural eutrophication of Lake Erie in the 1960's was the catalyst for scientific research into the causes of, and possible remedies for, this algal feeding frenzy.

It proved difficult, however, to study an entire ecosystem–its pollutants, how the pollutants react to each other, and how to stop the pollution–in the confines of a laboratory. Scientists determined it would take a natural ecosystem to serve as their laboratory. In 1969 Dr. David Schindler began a study of this very sort in the Experimental Lakes Area (ELA) of northwestern Ontario. Here, lakes that had been isolated from human activity leading to pollution served as a baseline to study the effects of nutrient loading. To determine the ultimate cause of cultural eutrophication, several lakes were loaded with different combinations of the same farm and lawn fertilizers.

Lake 227 was fertilized with both nitrogen and phosphorus. Before it was fertilized, the peak summer phytoplankton biomass was about 4-5 $g/m^3$ and consisted primarily of *Chrysophyta* (golden-brown algae). After only two years of fertilization the phytoplankton biomass was an alarming 35 g/m3, and the dominant species had changed to *Cyanobacteria* (blue-green algae) (Freedman, 1989). Researchers concluded that the nutrients had a significant impact on the growth of the biomass, but it was unclear which nutrient was causing the algal blooms.

To learn more, researchers conducted a similar study on Lake 226 from 1973 to 1983. This lake was divided into two lakes by a plastic partition. Nitrogen and carbon were added to one side of the lake, while nitrogen, carbon and phosphorus were added to the other. Within a few weeks the lake fertilized with nitrogen, carbon and phosphorus grew a thick algal mat on the surface. The lake that had been fertilized with only nitrogen and carbon remained essentially unchanged.

*Algal bloom on a lake caused by phosphates. American Water Works Association.*

This study clearly implicated phosphorus loading as the primary cause of cultural eutrophication in similar glacial lakes, such as Lake Erie. Recovery to near prefertilization levels was very rapid when phosphate additions were discontinued (Wetzel, 2001, p. 277).

This study and others at the ELA have led to voluntarily removal of phosphates from laundry detergents in the United States (1994) and Canada. Also wastewater treatment technology has evolved to incorporate phosphate removal (Litke, 2001).

We now know that both nitrogen and phosphorus can cause cultural eutrophication, depending on conditions. Applying the N/P ratio can determine which is the limiting nutrient in a particular system.

### Test Kit Activities

Place 5 clean funnels in a rack, and line each with filter paper (plastic 2-liter bottles can be modified to serve as funnels, with an upside down cardboard box with holes cut in it to serve as a rack). Place a

small container under each funnel to capture the sample water. Fill each funnel to a depth of one-third with soil (not sod). Add nothing to the control sample. Add the same amount of fertilizer to each of the additional funnels as follows: to funnel 1 apply solid commercial fertilizer (e.g., Miracle Grow®, Shultz®); to funnel 2 apply the same amount of liquid commercial fertilizer (e.g., Miracle Grow®, Shultz®); to funnel 3 apply a commercial steer manure to the surface; and to funnel 4 apply plant residue (e.g., compost, leaves, grass clippings, etc.).

Water each sample, including the control, with the same known volume of water. Test the water that has percolated through the soil and into the collection container for nitrates and phosphates. Record results in a data table. Repeat the following day, and for several days. Does one nutrient leach through the soil better than the other? Graph the results to illustrate how nutrients move through soil over time.

**References**

American Chemical Society. 2002. *ChemCom: Chemistry in the Community.* New York, NY: W.H. Freeman and Company.

Arms, K. and P. Camp. 1979. *Biology.* New York, NY: Holt, Rinehart & Winston.

Bartenhagen, K, M. Turner, and D. Osmond. 1995. *Phosphorus.* Retrieved on October 29, 2001 from the North Carolina State University WATERSHEDDS: Water, Soil and Hydro Environmental Decision Support System web site: http://h2osparc.wq.ncsu.edu

Freedman, B. 1989. *Environmental Ecology.* San Diego, CA: Academic Press, Inc.

Hach Company. 2001. *Important Water Quality Factors.* Retrieved on October 2, 2001 from the Hach Company web site: http://www.hach.com/h20u h2wtrqual.htm

Hartman, L. and M. Burk. 2000. *Volunteer Stream Monitoring Training Manual.* Indianapolis, IN: Hoosier Riverwatch. http://www.HosierRiverwatch.com

Havlin, J., J. Beaton, S. Tisdale, and W. Nelson. 1999. *Soil Fertility and Fertilizers: An Introduction to Nutrient Management, 6th Edition.* Upper Saddle River, NJ: Prentice Hall.

Knochel, J., ed. 2001. *Phosphorus.* Retrieved on October 29, 2001 from The Linus Pauling Institute web site: http://lpi.orst.edu/infocenter/minerals/phosphorus/phosphorus.html

Litke, D. 2001. *Review of Phosphorus Control Measures in the United States and their Effects on Water Quality.* Retrieved on December 12, 2001 from the United States Geologic Survey web site: http://water.usgs.gov/nawqa/nutrients/pubs/wri99-4007/

Mitchell, M. and W. Stapp. 1997. *Field Manual for Water Quality Monitoring: An Environmental Education Program for Schools.* Dubuque, IA: Kendall/Hunt Publishing Co.

Physics Department, University of Coimbra, Portugal. n.d. *Phosphorus.* Retrived on October 29, 2001 from the web site: http://nautilus.fis.uc.pt/st2.5/scenes-e/elem/e01510.html

Reuther, C. 1998. Microscopic Murderer. Retrieved November 20, 2001 from the Academy of Natural Sciences web site: http://www.acnatsci.org/erd/ea/pfiester.html

United States Environmental Protection Agency. 2001. *What You Should Know about Pfiesteria piscicida.* Retrieved November 20, 2001, from the web site: http://www.epa.gov/owow/estuaries/pfiesteria/fact.html

United States Environmental Protection Agency. 1997. *Volunteer Stream Monitoring: A Methods Manual.* United States Environmental Protection Agency.

Welch, E. B. 1980. *Ecological Effects of Waste Water.* Cambridge, UK: Cambridge University Press.

Wetzel, R. 2001. *Limnology, 3rd Edition.* San Diego, CA: Academic Press.

2000. *The American Heritage Dictionary.* Houghton Mifflin Company. Retrieved on November 20, 2002, from the web site: http://www.dictionary.com

# Temperature

## Primary Importance

Temperature dictates metabolic function, reproductive timing and duration, and, therefore, the life cycle of aquatic organisms. Temperature also affects other water quality parameters such as dissolved oxygen.

## Technical Overview And History

Because it takes so much to affect water temperature, changes in temperature occur very gradually in nature. Under natural conditions, water temperature changes much more gradually than air or land temperatures do. For example, although air temperature may change by 20° C in a 24-hour period, water temperature will change insignificantly during that time. Any increase in water temperature reflects a significant transfer of heat energy.

| Table of Relevant Specific Heats (Ebbing, 1990, p. 200) | |
|---|---|
| **Substance** | **Specific Heat in Joules (g·°C)** |
| Aluminum, Al | 0.901 |
| Copper, Cu | 0.384 |
| Iron, Fe | 0.449 |
| Calcium Carbonate, $CaCO_3$ | 0.818 |
| Water, $H_2O$ | 4.18 |

**Sources of Temperature Change**

Natural
- Sediment
- Vegetation along stream
- Surface area
- Composition of river bed
- Stream flow/volume
- Tributary temperature

Human
- Thermal pollution
- Runoff
- Removal of vegetation

**Relevance**

Animals and plants
- Affects dissolved oxygen
- Controls sensitivity to disease, pathogens and parasites
- Controls body temperature
- Affects life cycle (timing, duration and success)
- Controls rate of photosynthesis

**Impacts**

Environmental health
- Disrupts metabolic function in cold-blooded organisms
- Reduces dissolved oxygen
- Increases plant and algal growth
- Species shift (to more pollution tolerant species)

**Possible Remedies**
- Maintain vegetation along stream

The concept of specific heat is critical to understanding water temperature measurement. Specific heat is the amount of heat per unit mass required to raise the temperature of a material by one degree Celsius (°C) (Ebbing, 1990). It takes a significant amount of heat to raise the temperature of a unit mass of water 1° C (Wetzel, 1983).

So why does water ($H_2O$) have this uniquely high specific heat? Imagine the water molecule as a Mickey Mouse head, with the ears as hydrogen atoms and the face as an oxygen atom. The hydrogen ears have a slight positive charge, and the chin of the oxygen face has a slight negative charge, making water a polar molecule. The positively charged hydrogen atoms are attracted to negatively charged oxygen atoms of nearby water molecules. This particular attraction of one water molecule to a neighboring water molecule is called a hydrogen bond.

The hydrogen bond, although not a true chemical bond, is tight because the small hydrogen atoms are able to get very close to the oxygen atom of the nearby water molecule.

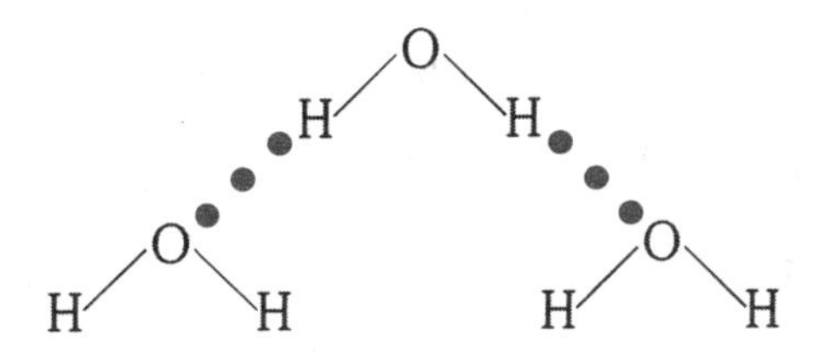


*Hydrogen bonding (represented by dots) between water molecules*

It takes a lot of heat to break those hydrogen bonds. When they are broken, the water molecules can move faster, and the manifestation of faster moving molecules of a substance is an increase in temperature of that substance.

### Sources

Natural physical factors, often modified by human intervention, can control the rate and degree of water temperature change. These factors include amount of vegetation along stream banks, surface area of a water body, river bed composition, streamflow or volume, water velocity, air temperature and the contribution of springs and tributaries.

Vegetation along a river affects temperature in two ways. First, it provides shade. Removing streamside vegetation can increase the temperature of the water by as much as 4° C (Freedman, 1989). Second it reduces the amount of sediment that enters a river via erosion and runoff. Sediment, which makes a river cloudier or more turbid, absorbs solar radiation and raises the water temperature.

Increasing the surface area of a water body causes the temperature to rise for a similar reason. When a larger area of water is exposed to the sun's powerful radiation, it absorbs more energy and the water temperature rises. The surface area of a water body increases when it is dammed to create a reservoir; when sedimentation increases, or when the channel width increases.

The composition of a river or lake bed also affects the temperature of the water. Dark materials absorb solar energy while lighter materials are more reflective. Also solid rock will absorb more of the sun's radiation than if it were gravel, sand, or silt.

The volume of a water body also influences its temperature. Shallow waters warm faster than deeper waters and slow-moving waters warm faster than swift waters. Shallow, slow-moving waters can also be susceptible to diurnal temperature changes.

Runoff also affects water temperature. Impermeable surfaces such as sidewalks, rooftops, parking lots, and streets absorb the sun's radiation and pass this heat to water that flows over them. This phenomenon is particularly significant in storm drain effluent from urban areas.

*Aquatic invertebrates are affected by changes in water temperature. Courtesy: Project WET*

### Relevance

Temperature is an important water quality parameter because it controls other water quality parameters such as dissolved oxygen, the rate of photosynthesis of aquatic plants and algae, the sensitivity of aquatic organisms to disease, pathogens, and parasites; as well as the body temperature and life cycle of most aquatic organisms.

Most aquatic organisms are cold-blooded, meaning that they cannot internally control their body temperature. Instead they take on the temperature of the water around them. Water temperature controls the metabolic rates and life functions of these animals. Such aquatic organisms have established themselves where the temperature range allows efficient metabolic functions, successful reproduction, and survival of their eggs, larvae, and fry. The extreme lows and highs of this temperature range are called lethal limits. Near and beyond the lethal limits, viability is uncertain.

Changes in water temperature affect other critical water quality parameters. In general, as the temperature of water increases, so does the solubility of solids, while the solubility of gasses decreases (K.R.A.M.P., 1997). As the water temperature rises, dissolved oxygen levels decrease. As a general rule colder water can hold more dissolved oxygen, therefore supporting more aquatic organisms than water with less dissolved oxygen.

Warmer waters, on the other hand

*Shade from overhanging trees helps to keep streams cool. Courtesy: United States Environmental Protection Agency.*

can support more plant and algal life. The rate of photosynthesis increases with the rise in temperature, until about 89° F (32° C), above which the rate of photosynthesis decreases. Also, "bacteria and other disease-causing organisms grow faster in warmer water" (KanCRN, 1999, p.1).

### Impacts

Aquatic organisms have adapted to the gradual temperature changes in natural water systems. However, rapid temperature changes are detrimental to fish and macroinvertebrates. These changes disrupt metabolic function, reduce available dissolved oxygen, and make the aquatic animals more susceptible to disease and predators (Bartenhagen et. al., 1995).

Further, when the temperature rises, plant and algal growth increases, in turn blocking sunlight and killing plants deeper in the water column. The increased mass of dying plant matter then feeds oxygen-consuming bacteria. Eventually, either the dissolved oxygen levels are too low to support the existing life or the lethal temperature limits are exausted for fish, eggs, larvae, fry, and macroinvertebrates.

With a sustained increase in water temperature, cold-water species such as trout and stonefly nymphs are replaced by carp and dragonfly nymphs (Mitchell et al., 1997) and diatoms and green algae are replaced by *Cyanobacteria*, or blue-green algae.

### Possible Remedies

There are several ways to reduce the impacts of thermal pollution and water temperature increases. First, a buffer strip of vegetation along the banks of streams or shores of lakes offers beneficial shade and reduces erosion and sedimentation. Industries can monitor and control the temperature of their discharges. When a river must be engineered into a man-made channel, factors such as a light-colored channel bottom and a narrow width (reduced surface area) can help minimize temperature increases.

WWW.

### Case Study

Solar radiation is the primary source of energy for increasing water temperatures in rivers and lakes. Much of this warming is natural and beneficial to fish and other aquatic organisms that have adapted to specific temperature ranges and the associated habitats. However, studies from diverse climates across the United States and Canada reveal that removing vegetation from streams and rivers results in water temperatures that exceed natural levels, often harming and sometimes killing fish and aquatic organisms (see Impacts).

One study in Oregon measured water temperatures from two years prior to removal of vegetation through four years after the removal. Three adjacent watersheds of the Oregon coast range were studied. One watershed had vegetation completely removed; one had vegetation removed to within 100 feet from the stream leaving a vegetative buffer strip; and a third watershed remained intact as a control (Brown et al., 1970). In the watershed with the complete vegetation removal, researchers found the mean monthly maximum temperature in July increased from 57° F to 71° F (a 14° change) the

year following removal. They found that the watershed with the 100-foot buffer strip showed no significant stream temperature changes. The control watershed provided a basis for seasonal variations in stream temperatures.

Vegetative buffer strips are an effective method of maintaining natural stream temperatures. Vegetative buffer strips also reduce erosion and runoff and provide woody debris and detritus that serve as nutrient and habitat sources for aquatic organisms (EPA, 1999). Landowner participation is essential as large strips of vegetation could reduce acres and farm or ranch production.

Vegetative buffer strips with a minimum width of 35 feet, can conserve the aquatic ecosystem, preserve natural water temperatures, and control erosion rates (EPA, 1999).

### Test Kit Activities

Fill at least two containers (same shape, size, and color) to two-thirds depth with tap water. Measure the temperature of all samples to confirm that they are starting at the same temperature. Leave one sample as a control, and add sediment to the other samples, mixing them well. You may wish to add different kinds and quantities of sediment to other containers. Place them uncovered in a sunny location, and record their temperatures throughout the day. How do the temperatures vary between samples? How might turbidity affect water temperatures in nature?

As an extension to this activity, you may wish to record the dissolved oxygen level of the water samples at the beginning of the experiment, when all samples are the same temperature, and at the end of the experiment, when samples have reached different temperatures. How does water temperature influence dissolved oxygen levels?

### References


Bartenhagen, K, M. Turner, and D. Osmond. 1995. *Temperature.* Retrieved on October 29, 2001 from the North Carolina State University WATERSHEDDS: Water, Soil and Hydro Environmental Decision Support System web site: http://h2osparc.wq.ncsu.edu

Brown, G. and J. Krygier. 1970. Effects of Clear-Cutting on Stream Temperature. *Water Resources Research.* 6 (4): 1133-1139.

Ebbing, D. 1990. *General Chemistry, 3rd Ed.* Boston, MA: Houghton Mifflin Co.

Environmental Protection Agency. 1999. *Streamside Management Areas.* Retrieved on April 23, 2002, from the website: http://www.epa.gov/OWOW/NPS/MMGI/Chapter3/ch3-2b.html

Freedman, B. 1989. *Environmental Ecology.* San Diego, CA: Academic Press, Inc.

Mitchell, M. and W. Stapp. 1997. *Field Manual for Water Quality Monitoring: An Environmental Education Program for Schools.* Dubuque, IA: Kendall/Hunt Publishing Co.

Swift, L. and J. Messer. 1971. Forest Cuttings Raise Temperatures of Small Streams in the Southern Appalachians. *Journal of Soil and Water Conservation.* 26 (1): 111-115.

Welch, E. B. 1980. *Ecological Effects of Waste Water.* Cambridge, UK: Cambridge University Press.

Wetzel, R. G. 1983. *Limnology, 2nd Edition.* Philadelphia, PA: Saunders College Publishing.

1997. *Temperature and Water Quality.* Retrieved on October 2, 2001 from the Kentucky River Assessment Monitoring Project (KRAMP) web site: http://water.nr.state.ky.us/ww/ramp/default.htm

1999. *Water Quality Index: Temperature.* Retrieved on October 8, 2001 from the Kansas Collaborative Research Network (KanCRN) web site: http://www.kancrn.org/stream/


Notes:

# Total Dissolved Solids

### Primary Importance

Solids dissolved in water are often essential nutrients for plants and animals.

### Technical Overview and History

Water is often referred to as the universal solvent because it is able to dissolve so many salts, minerals, and elements. As water runs over soil and rock or reacts with the atmosphere and the biosphere it dissolves many of the salts, minerals, elements, and organic compounds it comes in contact with and carries them away. Compounds such as inorganic salts that ionize when dissolved in water contribute to the total dissolved solids (TDS) level of the water. The ions, or molecules with a positive or negative charge, that commonly make up TDS include: calcium ($Ca^{2+}$), magnesium ($Mg^{2+}$), sodium ($Na^{+}$), potassium ($K^{+}$), sulfate ($SO_4^{2-}$), phosphate ($PO_4^{3-}$), nitrate ($NO_3^{-}$), carbonate ($CO_3^{2-}$), bicarbonate ($HCO_3^{-}$), and chloride ($Cl^{-}$).

Total dissolved solids are not an enforceable drinking water standard. Instead, the Environmental Protection Agency (EPA) has designated TDS as a secondary drinking water standard, meaning it may affect the cosmetics (i.e. appearance) or aesthetics (i.e. taste) of a water, but is not harmful to human health.

**Sources**

Natural
- rocks and soil
- decayed plants/animals
- erosion

Human
- urban runoff
- industrial effluent
- sewage effluent
- erosion

**Relevance**

Human
- better indicator of pollution than conductivity
- influences taste of drinking water

Animals and Plants
- regulates flow of water to cells
- can be nutrients necessary for life

**Impacts**

Human health
- high levels cause laxative effect
- unpleasant taste of drinking water if too high or low

Environmental health
- water balance problems in cells if high or low
- low levels inhibit growth of aquatic organisms

**Possible Remedies**
- reduce erosion
- maintain vegetation along waterways

However, constituents such as nitrates and phosphates that contribute to TDS levels may, individually, be regulated by the EPA.

Dissolved solids are traditionally measured gravimetrically. In this technique, a known sample is evaporated to dryness and the resulting precipitate weighed on a balance. Conductivity or specific conductance, which can be accurately and easily measured in the field, is often used as a quick and convenient estimate of TDS. The conductivity reading frequently is used to monitor water quality; it can provide the signal of a serious water quality problem such as improperly treated sewage or a water security threat.

### Sources

The local geology greatly influences the quantity and type of dissolved solids in the water. For example, waters that encounter limestone will have a relatively high level of TDS, because limestone dissolves easily when in contact with water. In this case, TDS will consist of calcium, carbonate, and bicarbonate.

**Limestone dissociates in water to create ions:**

$$CaCO_3 \rightarrow Ca^{2+}\ CO_3^{2-}$$

On the contrary, granite does not dissolve easily. Therefore, water that is in contact with granite or

igneous rocks will have a relatively low level of TDS.

Plants and animals contribute organic particles to soil and water as they die and decay, increasing the TDS. This contribution tends to vary with the seasons. At the onset of spring, snowmelt flushes out the decayed matter stored in soil and carries it over land in runoff to streams, rivers, lakes, and oceans. Therefore, very early in the spring TDS levels often are elevated. Once spring runoff has peaked, the soil has been sufficiently flushed, and the increased volume of receiving water results in diluted ion concentrations and decreased TDS levels. The seasonal change in TDS requires water treatment plants to adjust treatment strategies based on the amount and composition of TDS.

Similarly, non-seasonal erosion–either natural or anthropogenic–increases TDS when it transports soil constituents to water, where they dissolve. Activities that increase erosion include forest fires, clear-cut logging, mining, construction, and agriculture.

Agricultural and home lawn fertilizing contribute nitrates and phosphates that add to TDS in water. Lawn fertilizers often enter waters via urban runoff, which may also carry salts from winter de-icers, oils and other auto fluids. Human waste also contributes to TDS in the form of nitrates, phosphates and organic matter because "it is impossible for [wastewater treatment plants] to remove all of the dissolved solids contributed to the sanitary sewer" (Murphy, 2001). Typically, the more times water is removed from a natural source, used by humans, and returned to a water source, the higher the TDS.

### Relevance

Many of the ions that make up TDS combine with other ions and atoms to form molecules that are necessary for life. For example, phosphorus from phosphates is a component of DNA and RNA.

In addition, the level of total solids in water, which includes both dissolved and suspended solids, dictates the rate of osmotic balance or water flow in and out of cells. A delicate balance must be maintained in order to facilitate a healthy flow and thus perpetuate aquatic life.

It is apparent that certain levels of TDS make drinking water taste better to humans. A panel of testers determined the palatability of water as related to TDS in the following table (World Health Organization, 1996):

| TDS (mg/l) | Rating |
|---|---|
| Very low | unacceptable, flat, insipid taste |
| < 300 | excellent |
| 300-600 | good |
| 600-900 | fair |
| 900-1200 | poor |
| > 1200 | unacceptable |

### Impacts

The level of TDS in drinking water affects the taste or palatability. High levels of TDS can cause the water to have an unpleasant mineral taste and may have a laxative effect on the consumer. Very low levels cause the water to taste flat. High levels of TDS cause scaling in boilers and pipes, while water with low levels of TDS may corrode boilers and pipes. Both extremes may decrease the life of the boiler-system equipment.

Further, low concentrations of TDS may impede the growth of aquatic organisms that depend on the dissolved nutrients in water. For example, phytoplankton survive on nitrates and phosphates dissolved in water; without these constituents, phytoplankton will cease to grow.

### Possible Remedies

A moderate level of TDS is healthy and desirable for both the survival of aquatic organisms and the palatability of drinking water for humans. Therefore, a remedy is not to eliminate all sources of TDS, but rather to maintain a healthy balance in the watershed.

Proper management of all anthropogenic sources of ions can remedy virtually any ion imbalance in waterways. Proper management entails low-erosion farming practices, maintaining a buffer strip of vegetation along waterways to reduce erosion, minimal use of chemical de-icers in the winter, safe disposal and correct use of house-

*Lined canal reduces total dissolved solids. Courtesy: Gallatin, MT Conservation District: Natural Resources Conservation Service.*

hold chemicals and fertilizers, and thorough treatment of industrial and sewage effluent.

### Case Study

A major environmental issue facing the Colorado River is its increasing level of total dissolved solids, mostly in the form of dissolved salts (salinity). Near its headwaters, the salinity levels of the Colorado River are relatively low. The igneous and metamorphic rocks of Rocky Mountain National Park, where the river originates, do not erode easily and contribute very little salinity to the river. According to the United States Geological Survey, salinity near the headwaters is generally less than 50 mg/l (ppm) (Hart, et. al, 2000).

As the river flows to the Gulf of Mexico, the salinity concentration increases. Roughly half of the increase comes from natural sources, which includes geology, evaporation, and influent springs (Pontius, 1997). Downstream from the headwaters, the river eventually flows over sedimentary rocks, dissolving salts and picking up sediment, thus increasing the total dissolved solids and conductivity. Since the river flows through an arid region, water evaporates from its surface and the volume of water decreases as the concentration of salt and sediments in the water increases. Finally, mineral springs in the Colorado River watershed contribute salts and other minerals to its waters. In fact, "natural hot springs contribute about 500,000 tons of dissolved solids annually to the streams in the basin" (Apodaca, et. al, 1996, p. 2).

The other half of the Colorado River salinity comes from anthropogenic sources, such as municipal and industrial effluents, reservoir evaporation, use by introduced (non-native) phreatophytes (water-loving plants, such as willows and tamarisk), and irrigation return flow. High salinity levels corrode pipes by disrupting protective scale formation, thus increasing the cost of drinking water treatment. Irrigation water with high salinity decreases crop yields and requires greater volumes of water to dilute the build-up of salts (Pontius, 1997). Phreatophytes concentrate salts as they use water for growth and leave salts behind.

For example, "In 1964, salinity became an international issue when the Mexican government complained that deliveries of Colorado River water with salt concentrations of 2,000 ppm were affecting [farmers'] ability to grow crops and asserted that this was in violation of the 1944 Mexican Water Treaty" (Pontius, 1997, p. 60). This Treaty secured a quantity of water for Mexico from the Colorado River, but did not address water quality. These assertions precipitated a dialogue that lead to Minute No. 242 to the Treaty.

Minute 242 of the Mexican Water Treaty was signed in 1974, and the Colorado River Basin Salinity Control Act (CRBSCA) was passed to implement the Treaty. Both

Title I and Title II of the CRBSCA authorize measures to control the salinity of the Colorado River. Under Title I, construction projects initiated included the Yuma Desalting Plant, a canal to bypass nearby irrigation waters, and a well field to supplement river flows. In addition, pumping of water near the border was restricted, 10,000 acres of farmland was retired, and canals were lined to reduce seepage. Title II designated four salinity control units, or regions, within the Colorado River Basin. These units are used to conduct research on reducing or minimizing salinity in the River.

The basin-wide research program and subsequent salinity control measures have been so successful that the Yuma Desalting Plant is no longer in operation. While Minute 242 and the CRBSCA were precipitated by salinity levels of 2,000 ppm, salinity levels at the border between the United States and Mexico now are about 900 ppm (Hart, et. al, 2000). Additional salinity control research and implementation has been, and will continue to be, conducted in the Colorado River watershed.

**Test Kit Activities**

Collect soil samples from various locations, such as agricultural land, wetlands, forest, garden, playground, etc. Break up large chunks, and spread soil on a table, desk, or countertop to air dry. Be sure to label samples, and prevent cross-contamination by separating the samples. Drying should take several days, depending on humidity levels and the initial moisture content of the soil.

Next, place about 20 grams of dry soil from each sample in separate beakers and label the beakers accordingly. Add 20 ml of deionized water to each beaker. Stir soil and water samples for one minute at 10-minute intervals, over a thirty-minute period. Measure the conductivity of each sample. This is an approximation of the soil salinity. Multiply the conductivity values by 0.67 for an estimate of the total dissolved solids in the soil water.

**References**


Apodaca, L., V. Stephens, and N. Driver. 1996. *What Affects Water Quality in the Upper Colorado River Basin.* Retrieved on April 19, 2002, from the United States Geological Survey website: http://webserver.cr.usgs.gov/Pubs/fs/fs109-96/pdf/fs109-96.pd

Hart, R. and R. Hooper. 2000. *Monitoring the Quality of the Nation's Largest Rivers.* Retrieved on April 19, 2002, from the United States Geological Survey website: http://water.usgs.gov/nasqan/progdocs/factsheets/clrdfact/clrdfact.html

Murphy, S. 2001. *Interpretation of Boulder Creek Watershed Total Dissolved Solids Data.* Retrieved on December 4, 2001, from the BASIN web site: http://bcn.boulder.co.us/basin/data/NUTRIENTS/bc/TDS.html

Pontius, D. 1997. *Colorado River Basin Study.* Retrieved on April15, 2002, from the Water in the West website: http://www.waterinthewest.org/reading/readingfiles/fedreportfiles.colorado.pdf

World Health Organization. 1996. *Guidelines for Drinking-Water Quality, 2nd Ed. Vol. 2. Health Criteria and Other Supporting Information.* Retrieved on December 4, 2001, from http://www.who.int/water_sanitation_health/GDWQ/Chemicals/tdsfull.htm

1999. *Water Quality Index: TDS.* Retrieved on October 8, 2001 from the Kansas Collaborative Research Network (KanCRN) web site: http://www.kancrn.org/stream/


Notes

# Turbidity/Transparency

## Primary Importance

Turbidity is a significant indicator of overall water quality for the aquatic community as well as human health. Erosion prevention is the best way to minimize turbidity in waterways and drinking water.

## Technical Overview And History

Although the terms "turbidity" and "transparency" are often used interchangeably, they are distinctly different water quality measurements. "Turbidity is the cloudy appearance of water caused by the presence of suspended and colloidal matter. Technically, turbidity is an optical property of the water based on the amount of light reflected by suspended particles" (Environmental Protection Agency, EPA, 1999). Very turbid water appears murky or cloudy because light is scattered by suspended and colloidal matter such as sediment, minerals, microorganisms, and chemicals.

Transparency measures the clarity, or clearness, of water and is an indicator of how well light passes through it. Transparency can be affected by color (tannins) as well as by suspended materials. However, the impact of color on transparency doesn't necessarily represent turbidity.

There are several methods available for measuring transparency and turbidity. A secchi disk, (a specially marked plate 20 cm in diameter) is used to measure transparency in slow moving, deep water. The disk is divided into four equal quadrants in an alternating pattern of white and black. The user lowers the disk, attached to a cord, into the water. The depth at which the disk becomes invisible is known as the Secchi Depth and is recorded in meters (m) (Murdoch et al., 1991). A shallow Secchi depth indicates relatively unclear water, while a deeper Secchi depth indicates clearer water.

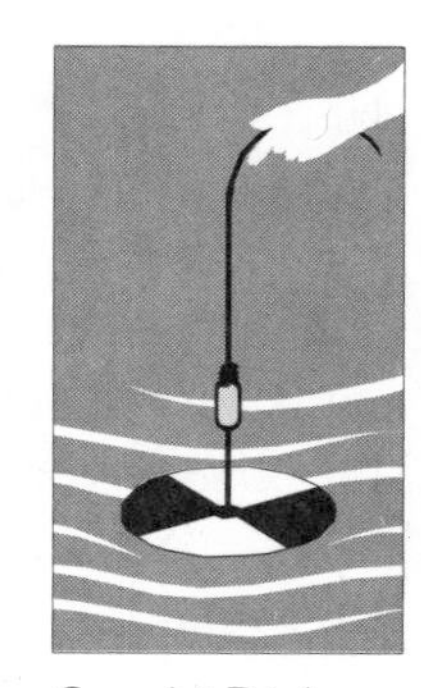
*Secchi Disk*

A transparency tube is a tall, clear plastic tube containing a Secchi disk pattern at the bottom and a centimeter scale from the bottom to the top of the tube. A water sample is poured into the tube, and water is released from the bottom (via a small outlet) until the Secchi disk pattern is visible.

To measure turbidity of a fast-moving, shallow stream, testers collect discrete samples to measure in a nephelometer, or turbidimeter. This device uses a light beam and a photoelectric cell to electronically measure, in quantitative Nephelometric Turbidity Units (NTU's), the amount of light scattered by suspended particles (www.hach.com). The higher the NTU (turbidity) value, the more light is scattered and, generally, the more solids it contains. However, because different colors and sizes of particles reflect light differently,

### Sources of Turbidity

Natural
- Sand, silt, & clay
- Algae, plankton, tannins
- Bacteria & viruses

Human
- Industrial waste
- Sewage
- Asbestos, lead, etc.

### Relevance

Human
- Impairs drinking water
- Impairs manufacturing of beverages, computer chips, etc.

Animals
- Smothers gravel beds
- Hides fish prey

Plants
- Carries nutrients

### Impacts

Human health
- Impairs drinking water disinfection
- Damages boilers and pipes

Environmental health
- Interferes with fish respiration and survival
- Decreases photosynthesis

### Possible Remedies

- Retain a buffer strip of vegetation along stream banks
- Proper treatment of sewage
- Contour farming

turbidity cannot be considered a direct measure of the quantity of total suspended solids in a sample.

Historically, the Jackson Candle was used to measure high levels of turbidity. Here a candle is lit beneath a long, clear glass tube and a water sample is poured into the tube until the observer, looking down the tube, can no longer see the candle through the sample.

### Sources

What makes up "suspended and colloidal matter"? Suspended solids can be anything suspended in the water column, from sand, silt, and clay, to algae and plankton, industrial wastes, sewage, lead, asbestos, bacteria, and viruses. Some suspended matter naturally occurs, and some is a product of human activities. Some suspended matter is visible, some is microscopic.

Sediment is created from the natural process of erosion, where wind, water, frost and ice slowly break rocks into finer and finer pieces.

*Erosion on a bare hillside. Courtesy: Natural Resurces Conservation Service*

This annual cycle, rain, snow, and snow melt causes natural variability in turbidity. Some of this sediment is carried over land by runoff to streams, rivers, lakes, and oceans. Fires, spring runoff, flooding and other natural events can temporarily elevate the level of suspended solids and turbidity.

Agriculture, construction, logging, and other vegetation removal can also increase erosion. Urban runoff also carries suspended matter into water systems. Poorly functioning wastewater treatment plants can release nutrients, which promote increased algal growth and plankton production.

### Relevance

Naturally-occurring turbidity often contains nutrient-laden sediments, which are necessary for plant growth. However, excessive turbidity may adversely affect plants, fish and macroinvertebrates (see Impacts).

The proper disinfection of a municipal water supply requires filtering of suspended solids to very low levels. The manufacture of paper, all beverages, and integrated circuits (computer chips) requires water free of suspended solids. In all of these processes, turbidity is a significant control parameter.

To avoid a shift in the aquatic species of a watershed, or a waterborne disease outbreak, sediments must be managed (Farthing et al., 1992, p. 132). Turbidity is the most standardized and easiest-to-implement way to assess particles and sediment in water. Turbidity is such a basic indicator of drinking water quality that it "is a parameter that the Environmental Protection Agency (EPA) requires to be tested on a daily basis" (Everpure, 2001).

### Impacts

Excessive turbidity impedes aquatic plant photosynthesis by blocking sunlight. Suspended particles actually absorb sunlight, causing water temperature to rise and dissolved oxygen levels to fall. The plants that die as a consequence become food for bacteria that, in turn, consume more dissolved oxygen.

In headwater streams especially, high levels of suspended solids in the water settle out and deposit in the spaces or interstices between rocks. This causes many problems for animal life. First, valuable habitat for macroinvertebrates, fish eggs, and fry is destroyed. Eggs need clean, oxygenated water for survival. Sediment deposition can suffocate these eggs and emerging fry, and cover up food sources for bottom feeders. It can choke filter feeders, and clog the gills of fish, thus reducing their ability to feed and decreasing their resistance to disease. High turbidity can also prevent certain fish from spotting their prey.

Humans are also affected by excessive turbidity. Suspended solids can harbor bacteria, protozoa and viruses, thus reducing the effectiveness of chemical disinfection (chlorination) of drinking water. Turbid

waters can also damage boilers and foul pipes.

### Possible Remedies

The best way to protect a stream's gravel bed is to stem sediment flow into our waterways Sediments will naturally settle to the bottom of rivers and lakes when streamflow or lake current velocity decreases. Erosion control and slower runoff are critical. Plant cover reduces soil erosion and runoff.

Retaining a buffer strip of plants along waterways will reduce bank erosion. Vegetation prevents direct impact of raindrops on the soil and keeps finer, lighter-weight soil particles from being forced to the soil surface. When finer soil particles collect at the surface, they can create a barrier that prevents water absorption and encourages runoff (Farthing et al., 1992).

Roots of plants act as conduits that direct water into the ground and slow run off. Fallen leaves and stems slow runoff and allow the water to soak into the ground. Finally, plants break down, and add organic matter to the soil that decomposes to water-absorbing humus.

*Turbidity in the Chattahoochee River, GA at the mouth of James Creek. Courtesy: Joe and Monica Cook.*

The effects of agriculture can be reduced by using contour cultivation methods and by retaining crop stubble (Farthing et al., 1992; Melbourne Parks and Waterways, 1995). Contour farming involves planting crops across a hillside in rows that mimic contour lines (lines of equal elevation). Leaving crop stubble or remnants during harvest also helps reduce erosion.

Proper filtration will remove most of the suspended solids from wastewater treatment plant effluent and drinking water treatment plant influent. Turbidity reduction during treatment can reduce nutrients, harmful bacteria, and viruses in both drinking water and receiving water.

### Case Study

Turbidity is a measure of the amount of light scattered by suspended particles in water such as sediment, minerals, microorganisms, and chemicals. It is widely recognized that the removal of vegetation to the edge of a stream, river, or lake increases the rate of erosion into that water and its turbidity (Packer, 1967b; Burns, 1972; Corbett et al., 1978; Murphy et al., 1981; all found in Freedman, 1989).

Studies indicate that removing vegetation from land does not have to cause such a dramatic increase in erosion rates (Tjaden et al., n.d.; EPAb, 1999; Freedman, 1989; and Arthur et al., 1998). Best Management Practices (BMPs) have been developed (based on clear-cutting and sedimentation research) to minimize the amount of erosion and sedimentation flux (change in the amount of sediment transported) that results from removing vegetation. Some of these BMPs for vegetation removal include:

- Maintaining a buffer strip of vegetation about 50 feet wide lining the edges of water sources.
- Constructing roads with a minimum grade that does not exceed 10 percent.
- Re-vegetating roads, skid trails and landings after harvesting has ended.
- Minimizing compression of soils by using low-impact skidding practices (winching logs, varying skid trails, and avoiding downhill skidding).
- Cross-draining of roads.

Researchers evaluated the effectiveness of these practices in an Eastern Kentucky study in which they compared sedimentation flux and water quality parameters in three watersheds. One watershed was left intact as a control. Another watershed was clear-cut using the above BMPs, while a third watershed was clear-cut without the use of BMPs.

The scientists began collecting data more than a year before cutting began, (1982) to determine background water quality parameter levels (Arthur et al., 1998). Monitoring continued for 10 years (through 1992). Before harvesting began, slight differences in sediment flux among the three watersheds were noted. These differences, however,

**Approximate Suspended Sediment in kilograms per hectare (kg/ha)**

| | Pre-harvest (1982) | 1990 |
|---|---|---|
| Non-BMP watershed | 120 | 300 |
| BMP watershed | 100 | 50 |
| Un-cut watershed | 60 | 20 |

(Arthur et-al., 1998)

were not significant compared to the differences observed during and after harvesting.

Researchers found that sediment flux from the non-BMP watershed was about two times greater than that from the BMP watershed during the timber harvest. For the year following the harvest, sediment flux from the non-BMP watershed was about one and a half times greater than that from the BMP watershed. Both clear-cut watersheds had greater sediment flux than the un-cut control watershed. Even in 1990 (the final year sediment data were recorded), data indicate that the non-BMP watershed exhibited a much greater sediment flux than either the BMP watershed or the un-cut watershed (see table).

The BMPs applied in this study reflected 1984 technology. Researchers suggested that the results from the study "provide conservative measures of the effectiveness of current BMP guidelines, which have evolved to include more stringent guidelines for road building" (Arthur et al., 1998, p. 493). Using current Best Management Practices for all types of vegetation removal within watersheds can further reduce erosion and help maintain a healthier ecosystem.

**Test Kit Activities**

Collect water samples from various sources, including a river, lake, pond, puddle and tap. Use a turbidity tube or colorimeter to measure the clarity of the water. Compare the relative clarities of the samples. Have students speculate as to the sources of turbidity in each of the samples. Conduct this activity over a year to monitor seasonal trends in turbidity. Or, students could measure the turbidity of the waters in a watershed from its headwaters to its mouth, and demonstrate how turbidity changes over the length of the watershed.

**References**


Arthur, M., G. Coltharp, and D. Brown, 1998. Effects of Best Management Practices on Forest Streamwater Quality in Eastern Kentucky. *Journal of the American Water Resources Association.* 34 (3): 481-495.

Farthing, P., B. Hastie, S. Weston, and D. Wolf. 1992. *The Stream Scene.* Portland, OR: Aquatic Education Program.

Freedman, B. 1989. *Environmental Ecology.* San Diego, CA: Academic Press, Inc.

Murdock, T. and M. Cheo. 1996. *Streamkeepers Field Guide.* Everett, WA: The Adopt-A-Stream Foundation.

Environmental Protection Agency (EPA)a. 1999. *Guidance Manual for Compliance with the Interim Enhanced Surface Water Treatment Rule: Turbidity Provisions.* Retrieved on December 3, 2001, from the website: http://www.epa.gov/safewater/mdbp/mdbptg.html

Environmental Protection Agency (EPA)b. 1999. *Streamside Management Areas.* Retrieved on April 23, 2002, from the website: http://www.epa.gov/OWOW/NPS/MMGI/Chapter3/ch3-2b.html

Tjaden, R. and G. Weber. n.d. *Riparian Buffer Management: Riparian Buffer Systems.* Retrieved on April 24, 2002 from the Maryland Cooperative Extension website: http://www.agnr.umd.edu/MCE/Publications/Publications.cfm?ID=17

1995. *Physical and Chemical Tests: Turbidity.* Retrieved on November 27, 2001 from the Melbourne Parks & Waterways web site: http://redtail.eou.edu/streamwatch/swm19.html

1999. *Water Quality Index: Turbidity.* Retrieved on October 8, 2001 from the Kansas Collaborative Research Network (KanCRN) web site: http://www.kancrn.org/stream/

2001. *Important Water Quality Factors.* Retrieved on September 4, 2001 from the Hach Company web site: http://www.hach.com/h20u/h2wtrqual.htm

2001. *Turbidity in Drinking Water.* Retrieved on November 27, 2001 from the Everpure, Inc. web site: http://www.everpure.com/WaterU/11-turbidity

# HEALTHY WATER HEALTHY PEOPLE

## Field Monitoring Guide
## Appendixes

# Appendixes

# Interpreting Your Water Quality Data

| Parameter | Your data indicate: | Which may come from: | Which may be caused by: | Additional Notes: |
|---|---|---|---|---|
| **Alkalinity:** measures the water's capacity to neutralize acids in (mg/l) or (mg/l $CaCO_3$) | Increase | Wastewater treatment plant effluent<br>Surrounding geology | • Residues from food substances and cleaning agents<br>• Weathering and erosion of limestone | Total alkalinity of seawater averages 116 mg/l<br>Total alkalinity of freshwater is often between 30 and 90 mg/l |
| | Decrease | Acid precipitation<br>Surrounding geology<br>Industrial effluent | • Burning of fossil fuels<br>• Weathering and erosion of granite or igneous rocks<br>• Low pH water consumes alkalinity | |
| **Bacteria:** measures Fecal Coliform, E. coli, or Enterococci to indicate potential fecal contamination of water in Colony Forming Units (CFU) or cells /100 ml of water | Increase | Warmblooded animal or human feces<br>Soil<br>Pulp and paper mill wastes | • Leaking or failing septic and sewer systems<br>• Sewer overflows, overloaded/malfunctioning waste water treatment plant<br>• Runoff from areas containing pet and animal waste<br>• Direct defecation of animals and birds/waterfowl in waterways<br>• Many coliforms are naturally found in soil<br>• The genus *Klebsiella* of the Fecal coliform group common (harmless) | EPA criteria for Bacteria in recreational waters<br>Fresh water:<br>E. coli levels shall not exceed...<br>• 236 cells or CFU per 100 ml of a single water sample, or<br>• 126 cells or CFU per 100 ml as a geometric mean from at least 4 water samples<br>Marine wate..<br>Enterococci levels shall not exceed...<br>• 35 cells or CFU per 100 ml as a geometric mean from at least 5 water samples equally spaced over a 30-day period |
| | Decrease | Bacteria-free influent water | • Springs, tributary, runoff or other clean water source that reduces the total number of coliforms per unit volume of water | |
| **Biochemical Oxygen Demand (BOD):** measures the amount of oxygen in the water consumed by aquatic organisms and chemical reactions in (mg/l) | Increase | Organic matter | • Microorganisms consume oxygen when decomposing animal/pet waste, leaves and woody debris, nutrients | |
| | Decrease | Aeration<br>Waste water treatment plant effluent | • Increases the rate of decomposition of organic and inorganic material<br>• Chlorine kills decomposers (microorganisms) | |
| **Conductivity:** measures the water's ability to conduct an electrical current in<br>cont. next page | Increase | Urban runoff<br>Surrounding geology<br>Temperature | • Chemical de-icers, salts<br>• Clay soils dissolve into ionic components<br>• Conductivity is higher in warmer water | Conductivity of common waters in µmho/cm<br>Deionized water: 0-1<br>Distilled water: 0.5-3 |

# Interpreting Your Water Quality Data/Cont.

| Parameter: | Your data indicate: | Which may come from: | Which may be caused by: | Additional Notes: |
|---|---|---|---|---|
| micromhos/cm or microsiemens/cm (µS/cm) | Increase | Wastewater treatment plant effluent | • Ions such as chloride, phosphate, and nitrate | Rivers in the US: 50-1500<br>Healthy streams: 150-500<br>Some industrial effluent: 10,000<br>Seawater: 50,000 |
| | | Mining operations | • Ions such as sulfate, copper, cadmium, or arsenic in mine drainage | |
| | Decrease | Agricultural runoff | • Ions such as nitrate, phosphate, and salts | |
| | | Surrounding geology | • Granite and igneous rocks often do not dissolve into ionic components | |
| | | Industrial effluent | • Oils, alcohols, sugar, and many hazardous organic compounds reduce the number of charged ions per unit volume of water | |
| | | Urban runoff | • Oils | |
| **Dissolved oxygen (DO):** measures the amount of oxygen dissolved in water in (mg/l) | Increase | Aeration | • Waterfalls, rapids, rocks, and turbines add oxygen to water | Dissolved oxygen levels required for aquatic organisms<br><br>0 Prolonged<br>1 exposure<br>2 lethal<br>3<br>4 Stressful<br>5 to most<br>6 aquatic organisms<br>7<br>8 Usually required<br>9 for growth<br>10 and activity<br><br>Note: DO levels can fluctuate greatly with depth, especially in lakes and reservoirs |
| | | Photosynthesis | • Plants give off oxygen | |
| | | Temperature | • Colder water is able to dissolve more oxygen than warmer water | |
| | Decrease | Respiration, decomposition, and chemical reactions | • Plants, animals, and microorganisms consume oxygen during these processes | |
| | | Turbidity | • Blocks sunlight from plants, decreasing photosynthesis and increasing decay and decomposition<br>• Causes water temperature to rise by increasing absorption of solar radiation | |
| | | Ground water influent | • Low in dissolved oxygen because it is not exposed to the atmosphere while underground | |
| | | Elevation above sea level | • As atmospheric pressure decreases, less oxygen is dissolved in the water | |
| | | Reservoir bottom-water influent | • Low in DO due to reservoir stratification | |
| | | Nutrients | • Fuel overgrowth of algae, which die and decompose | |
| **Hardness:** measures the concentration of dissolved minerals by measuring polyvalent cations in (mg/l of $CaCO_3$) | | Surrounding geology | • Weathering and erosion of limestone | Water hardness levels in mg/l of $CaCO_3$<br><br>0-60 = soft water<br>61-120 = moderately hard<br>121-180 = hard water<br>181 and greater = very hard |
| | | Waste water treatment plant effluent | • Residues from food substances and cleaning agents | |
| | | Mining operations | • Expose rocks containing calcium and magnesium | |
| | | Surrounding geology | • Weathering and erosion of granite or igneous rocks | |

# Interpreting Your Water Quality Data/Cont.

| Parameter: | Your data indicate: | Which may come from: | Which may be caused by: | Additional Notes: |
|---|---|---|---|---|
| **Nitrate:** measures the organic or fertilizer matter in water in (mg/l) | Increase | Nutrients | • Runoff from agricultural land, residential lawns, and golf courses | The EPA suggests that unpolluted waters shall contain less than 1 mg/l of nutrients |
| | | Human and animal wastes | • Wastewater treatment plant effluent, runoff from areas containing pet or animal waste<br>• Direct defecation of animals and birds/waterfowl in waterways | |
| | | Burning of fossil fuels | • Releases long-term store of nitrogen | |
| | Decrease | Nutrient-free influent water | • Springs, tributary, runoff or other clean water source that reduces the level of nutrients per unit volume of water | |
| | | Plant use | • Nitrates are used by aquatic organisms for growth | |
| **pH:** measures the hydrogen ion concentration or activity on a logarithmic scale (no units) | Increase | Photosynthesis | • Plants use carbon dioxide, which reacts with water to form carbonic acid | acidic ↑ 0 1 2 3 4 5 6 neutral 7 8 9 10 11 12 13 basic ↓ 14 |
| | | Mining operations | • Acid mine drainage | |
| | Decrease | Respiration | • Plants give off carbon dioxide, which reacts with water to form carbonic acid | |
| | | Surrounding vegetation | • Sphagnum moss and pine needles are slightly acidic (bogs, marshes, and pine forests)<br>• Decaying vegetation produces organic acids | |
| | | Burning fossil fuels | • Emissions react with the atmosphere to form acid precipitation | |
| **Phosphate:** Measures the organic or fertilizer matter in water | Increase | Surrounding geology | • The mineral apatite contains phosphates | The EPA suggests that greater than 0.1 mg/l of total phosphates stimulates plant growth to surpass natural eutrophication rates |
| | | Human and animal wastes | • Waste water treatment plant effluent, runoff from areas containing pet or animal waste<br>• Direct defecation of animals and birds/waterfowl in waterways | |
| | | Nutrients | • Runoff from agricultural land, residential lawns, and golf courses | |
| | Decrease | Nutrient-free influent water | • Springs, tributary, runoff or other clean water source that reduces the level of nutrients per unit volume of water | |
| | | Plant use | • Phosphates are used by aquatic organisms for growth | |
| | Increase | Weather/seasonal | • Water temperature varies with air temperature: | |
| **Temperature:** measures the average amount of heat in the water in degrees Fahrenheit (°F) or degrees Celsius (°C) | | Removal of vegetation | • Stream-bank vegetation provides shade and reduces runoff (turbidity) | Many plants, bass, crappie, bluegill, carp, sucker, caddisfly, larvae, many fish diseases at temperatures around 20°C (68°F) |
| | | Impoundments | • Impoundments increase the surface area of the water that is exposed to | |

# Interpreting Your Water Quality Data/Cont.

| Parameter: | Your data indicate: | Which may come from: | Which may be caused by: | Additional Notes: |
|---|---|---|---|---|
| **Temperature:** cont. | | | solar radiation | Some plant life, trout, walleye, northern pike, stonefly nymph, caddis fly larvae, water beetles, water striders, some fish diseases 14°C (57°F) |
| | | Thermal pollution | • Warm-water discharge from power plants and other cooling waters | |
| | | Urban runoff | • Water is heated by asphalt and pavement | |
| | Decrease | Cold water inflow | • Groundwater, tributary, springs, and reservoir bottom water can be colder than the receiving water | Few plants, trout, caddisfly larvae, stonefly nymph, mayfly nymph, few fish diseases ?°C (?°F) |
| | | Depth | • Seasonal stratification, especially in lakes | |
| **Total dissolved solids (TDS):** measures ions and particles that will pass through a filter with pores of about 2-4 microns (0.002-0.004 cm) in size in (mg/l) | Increase | Seasonal/weather | • Runoff from rain an snowmelt carries dissolved constituents | According to the World Health Organization, TDS of natural water sources varies greatly from less than 30 mg/l to as much as 6,000 mg/l |
| | | Erosion | • Agricultural, road-building, construction, and logging all increase erosion rates. | |
| | | Waste water treatment plant effluent | • Many dissolved constituents are not removed | |
| | | Surrounding geology | • Limestone or sedimentary rocks dissolve easily, releasing ions | |
| | | Organic matter | • Decaying plants and animals | |
| | | Urban runoff | • Chemical de-icers, salts, fertilizers (nutrients), and chemicals | |
| | Decrease | Surrounding Geology | • Granite and igneous rocks often do not dissolve easily | |
| **Turbidity:** measures either the clarity of water using a Secchi Disk, or the amount of light reflected by suspended particles using a Turbidimeter in either a Secchi Depth (meters) or a Turbidimeter reading in Nephelometric Turbidity Units (NTU) | Increase | Erosion | • Agriculture, road-building, construction, and logging all increase erosion rates | Typical groundwater has a turbidity of less than 1 NTU |
| | | Nutrients | • Increased algal growth | |
| | | Weather/seasonal | • Runoff from snowmelt or rainfall | |
| | | Urban runoff | • Salts, fertilizers (nutrients), chemicals, sediment<br>• Increases stream-bank erosion since pavement prevents slowseepage of water into the ground | |
| | | Forest fires/flooding | • Temporarily increases erosion and therefore turbidity | |
| | | Waste water treatment plant effluent | • Nutrients increase algal growth<br>• Carries dissolved and suspended solids | |
| | Decrease | Erosion prevention measures | • Stream-bank vegetation; Best Management Practices for agriculture, road-building, construction and logging | |

# Cross Reference Between Healthy Water, Healthy People Publications and Testing Kits

| Activity Topics | Healthy Water, Healthy People Water Quality Educators Guide | Healthy Water Healthy People Testing Kits that Illustrate the Topics | Healthy Water, Healthy People Testing Kit Manual |
|---|---|---|---|
| Scientific Method | Carts and Horses | All Testing Kits | All |
| Nonpoint Source Pollution | Footprints on the Sand | All Testing Kits | All |
| pH | From H to OH! | All Testing Kits | Alkalinity, pH Chapters Appendix, A; Interpreting Your Water Quality Data |
| Drinking Water Contamination | Get the Lead Out | Drinking Water Testing Kit<br>Classroom Drinking Water Testing Kit<br>Advanced Water Quality Testing Kit | Nitrate, Bacteria Chapters Appendix A; Interpreting Your Water Quality Data |
| Understanding Measurement/Parts-Per-Million | Grab a Gram | All Testing Kits | Appendix E; Metric Conversions |
| Ground Water | Going Underground | Drinking Water Testing Kit<br>Classroom Drinking Water Testing Kit<br>Advanced Water Quality Testing Kit | Nitrate Chapter Appendix A; Interpreting Your Water Quality Data |
| Accuracy and Precision | Hitting the Mark | All Testing Kits | Appendix A; Interpreting Your Water Quality Data |
| Solutions & Mixtures | It's Clear to Me! | All Testing Kits | All |
| Microbial Contamination | Life and Death Situation | Classroom Drinking Water Testing Kit | Bacteria Chapter Appendix A; Interpreting Your Water Quality Data |
| Microbial Contamination | Looks Aren't Everything | Classroom Drinking Water Testing Kit | Bacteria Chapter Appendix A; Interpreting Your Water Quality Data |
| Aquatic Invertebrate Sampling | Benthic Bugs and Bioassessment | MacroPac™<br>Turbidity Tube | Appendix A; Interpreting Your Water Quality Data |
| Concept Mapping | Mapping It Out; Water Quality Concept Mapping | All Testing Kits | All |
| Differing Views Of An Issue | Multiple Perspectives | All Testing Kits | Appendix A; Interpreting Your Water Quality Data |
| Watershed Restoration | Picking up the Pieces | First Step™ Testing Kit<br>Watershed Testing Kit<br>Rivers, Streams, Ponds, and Lakes Testing Kit<br>Advanced Water Quality Testing Kit<br>MacroPac™<br>Turbidity Tube | Appendix A; Interpreting Your Water Quality Data |

# Cross Reference Between Healthy Water, Healthy People Publications and Testing Kits

| Activity Topics | Healthy Water, Healthy People Water Quality Educators Guide | Healthy Water, Healthy People Testing Kits that Illustrate the Topics | Healthy Water, Healthy People Testing Kit Manual |
|---|---|---|---|
| Water Quality Standards | Setting the Standards | All Testing Kits | Appendix A; Interpreting Your Water Quality Data |
| Watershed Monitoring | A Snapshot in Time | First Step™ Testing Kit<br>Watershed Testing Kit<br>Rivers, Streams, Ponds, and Lakes Testing Kit<br>Advanced Water Quality Testing Kit<br>MacroPac™<br>Turbidity Tube | Appendix A; Interpreting Your Water Quality Data |
| Alkalinity | Stone Soup | Watershed Testing Kit<br>Rivers, Streams, Ponds, and Lakes Testing Kit<br>Advanced Water Quality Testing Kit | Alkalinity, pH Chapters |
| Hydrologic Cycle | Pollution - Take It Or Leave It! | All Testing Kits | All |
| Nonpoint Source Pollution | There is No Point to This Pollution! | All Testing Kits | All |
| Turbidity & Sediment | Turbidity or not Turbidity - That is the Question! | Rivers, Streams, Ponds, and Lakes Testing Kit<br>Turbidity Tube | Turbidity, Temperature Chapters |
| Water Treatment & Wastewater Treatment | Washing Water | First Step™ Testing Kit<br>Drinking Water Testing Kit<br>Classroom Drinking Water Testing Kit<br>Advanced Water Quality Testing Kit<br>Turbidity Tube | All |
| Water Quality Monitoring & Study Design | Water Quality Monitoring; From Design to Data | First Step™ Testing Kit<br>Watershed Testing Kit<br>Rivers, Streams, Ponds, and Lakes Testing Kit<br>Advanced Water Quality Testing Kit<br>MacroPac™<br>Turbidity Kit | Appendix A; Interpreting Your Water Quality Data |
| Internet Web Quest & Website Evaluation | A Tangled Web; Conducting Internet Research | N/A | All |
| Water Quality Effects On Plants & Animals | Water Quality Windows | First Step™ Testing Kit<br>Watershed Testing Kit<br>Rivers, Streams, Ponds, and Lakes Testing Kit<br>Advanced Water Quality Testing Kit<br>MacroPac™<br>Turbidity Kit | All |
| Aquatic Macroinvertebrates as Water Quality Indicators | Invertebrates as Indicators | First Step™ Testing Kit<br>Watershed Testing Kit<br>Rivers, Streams Ponds, and Lakes Testing Kit<br>Advanced Water Quality Testing Kit<br>MacroPac™<br>Turbidity Tube | All |

# A Brief History of Chemistry

## Up to the Discovery of the Water Molecule

Bruce J. Hach; Managing Director, Hach Scientific Foundation
The Hach Scientific Foundation Funds Chemistry Teacher Scholarships.

**Ancient Philosophers**

What are the most fundamental "elements"? What are the smallest, the most undividable particles that make up all matter? What is the nature of matter? During the classic age of philosophy, about 300 B.C., thinkers from China, the Middle East and the Greek Empire contemplated these questions. Their thoughts and ruminations were accomplished without the knowledge of modern science principles as we know them today.

Many conjectured the fundamental elements to be fire, ice, water and various metals. Some considered associating opposites, such as dry vs. wet, hot vs. cold. Some considered colors and numbers to be elements. Even planets and celestial bodies were at one time considered to be elements.

The Greek philosopher, Democritus (460-370 B.C.) sensed the smallest fragments might be invisible and bind together to form matter. He even called these fragments "atomos." This may have been an initial spark in the creation of the atomic theory.

Socrates, Plato and finally Aristotle were dismissive of one aspect of the "atomos" theory. Although they believed that indivisible particles existed, they would not believe that there was nothing between the particles. They wanted to voice a conclusive argument that all matter is substance and would not entertain exploring the impractical view of the possibility of particles existing in a vacuum. These philosophers wanted the time and space to explain many, many practical phenomena. Aristotle took a centrist position and declared there are but four "elements," Fire, Water, Earth and Air. These were accepted as the fundamental elements. This discussion created a sense of curiosity about the nature of matter; however, the acceptance was so overwhelming it became a form of dogma. The four elements were almost universally accepted for the next 1,500 years.

Other Greek mathematicians were intrigued with "atomos". They visualized geometric mathematical shapes and interfaces. Thales, a Greek philosopher, saw single "elements" and sought to prove this fundamental principal with water. No one visualized chemistry.

The philosophers were thinkers only; no real experiments or science application would ever take place. Yet, the nature of matter is the first question addressed by the scientific development.

**Alchemy (300 B.C. to 1600 A.D.)**

The early form of chemistry was alchemy. From Egyptian and Arab origins, the practice of alchemy goes back over 2000 years. Alchemy tied many things together. There were some legitimate chemical developments in dyes, glassmaking, ceramics, acids, and metals, but there was also a philosophical component, rationalizations and vague explanations on creation, religion, and even the occult. Many think of alchemy as "transmutation"–trying to formulate pure gold from lead and iron. One form of alchemy was "Philosophers' Stone"; a type of mercury and sulfur concoction that would be an elixir of life. Touching it would heal any deformity. It would serve as the true fountain of youth. Alchemists worked on the stone for hundreds of years, but to no avail.

Alchemy was so varied–there was fundamental chemistry, but also folklore, mysticism, magic and perhaps even fraud. Many an alchemist would try to formulate a proprietary potion, but then kept the recipe a secret, misrepresented it, or used it exploitatively. The alchemist did not lead into systematic scientific discovery. By the 1600's, science and chemistry separated itself from alchemy.

**Paracelsus (Swiss–1493)**

A Swiss physician, Paracelsus, had an interesting middle name, Bombast. It is through his writings and exhortations we begin to see a break from the intransigence of the Middle Ages to the rational enlightenment of the forthcoming renaissance. Paracelsus had formulated medical remedies using mercury, arsenic, sulfur and lanthanum. These remedies were successful, making Paracelsus an accomplished physician.

Paracelsus visualized a relationship between chemistry, medicine and pharmacy, and even life sciences. While he attacked the mystical side of alchemy, he did accept Aristotle's concept of the

four "elements" of: fire, earth, water, and earth, but thought they were too broad, too general and did not lend themselves to scientific inquiry.

In response, Paracelsus brought in three principles, his new elements of: salt, mercury and sulfur. He saw these as the "Tria Prima" – closer to the material of life and earth. This was a chemical relationship closer to true elements. They were "primal bodies" and they suggested the need to understand a broader range of substances. His writings were often caustic and sarcastic and certainly aggressive, yet Paraclelsus was a voice in the development of science and a voice during a time of otherwise little development.

**John Baptiest Van Helmont (Belgium 1577-1644)**
Van Helmont was a physician who continued chemical development with a much gentler personality than Paracelsus. More of a true chemist, Van Helmont emphasized experiments with control, definition, quantification, and proper documentation. Scales, balances and thermometers were improving. Better equipment led to better methodology.

Accepting air and water as elements that can't be altered or divided, Van Helmont's early work focused on chemical reactions and gas. In the early 1600's he combined alkali and sand and formed water glass. This led to experiments in evaporation and eventually eight different types of gases. These "spirits" or vapors Van Helmont named "gas." While he didn't collect gaseous substances, Van Helmont was the first to isolate carbon dioxide gas.

Rejecting Aristotle's four elements, Van Helmont was also critical of Paracelsus' three prima bodies. Yet through gas characterization, separations and such disciplined methodology, Van Helmont was an influence on the discipline of science methods. One contribution was his immediate teaching and influence of Robert Boyle.

**Robert Boyle (English 1627-1690)**
Boyle's chemical development continued with vapors and gases. In 1661, Boyle wrote his famous "Skeptical Chymist" which continued to deplore crude alchemy and press for a more rational integration of chemical discoveries. It would ultimately fit together like a "great piece of clockwork."

Always showing every respect for Van Helmont's work, Boyle raised the standard for experiment rigor, discipline, and quantification. Boyle defined himself as an "atomist." He calls the smallest undividable particle a "corpuscle" and attributed it with movement and spacing in different states. This is the first verbal description of an element and later a compound.

Boyle invented the vacuum pump and studied gases. Boyle's law refers to the characteristics of a gas and how the volume of gas varies inversely with pressure. In all of his experiments, Boyle was the first, although inadvertently, to separate oxygen and hydrogen.

By the end of the 17th century, Boyle had identified chemistry as an objective scientific discipline thereby segregating it from alchemy.

**Phlogiston Era (1700-1785)**
In the history of chemical development, the phlogiston era was a setback. Authors Johann Bechler and George Ernst Stahl were considered top Dutch and Prussian chemists. Both renowned and popular, they were highly respected and admired. Chemists from the early $18^{th}$ century received training in the alchemy arena. There was confusion on chemical reactions of gases and the principles of combustion. Sometimes old philosophies can be roused to explain reactions that aren't fully understood.

The phlogiston theory suggests there was a "force within" and when metal was heated and burned, the phlogiston was given off and combined with other parts of nature. In other words, the contained phlogiston somehow contributed to combustion. Yet phlogiston, this 'force," was neither a material, nor an element, nor a tangible, nor an earth. It was thought to be a chemical or physical property that was "yet to be discovered."

Phlogiston was properly defined in proper chemical language, and the best scientists of the day were devoted and loyal to the theory, even including the mid-$18^{th}$ century chemists that were doing (performing) experimental discovery.

There were two serious flaws to the phlogiston beliefs that drew investigation. When a metal is burned, weight

is gained rather than lost. Combustion in a confined space causes the air to contract, not expand. Since the phlogiston theory was bogus, it would never be "discovered." Yet unusual chemical reactions do take place during combustion. No one could identify these chemical reactions, this "fuel for the fire." Since there were no other theories of combustion to take its place, the phlogiston theory carried creditability and remained well accepted for over 75 years.

### The Discovery of Water

Chemistry is the study of elements and the predictability of their reactions when combined. Elements were being discovered in the 18th century. However, when a non-element or non-entity is interjected into research, it causes confusion and the discovery process is slowed down.

Many phlogiston chemists continued to experiment in an attempt to validate and confirm their theories. There was confusion about the chemical make-up of gases; were they inherent? Were gases compounds or elements? Air was considered a single element. These questions, plus emerging doubt about phlogiston and the increasing academic rivalry of European nations spurred chemical research to accelerate in the middle 1700s.

One of the phlogiston's greatest champions was an emminent gas research scientist, Joseph Priestly (English 1733-1804). His pneumatic trough and engineered equipment were superb for collecting and characterizing gas. In 1774, Priestly completed an experiment that had been done by others, heating mercury oxide. Pure silvery mercury was separated, but also emitted was a very "pure air." Priestly collected the air and characterized its properties. He discovered that this "pure air" could allow a flame to burn brighter, mice could live and when inhaled, it would make a person "peppy." Priestly called this air, "dephlogisticated air." Wilhelm Scheele also heated mercury oxide with the same results; he called his air "fire air." However, no one knew what this substance might be.

Into the fray stepped a suave scientist who eventually formulated the correct interpretation. Antoine Laurent Lavoisier. Noted scientist, Lavoisier had recently been elected to the French science academy, a major voice in dissemination of research hypothesis and experimental science.

In 1772, Lavoisier discovered that when burning sulfur or phosphorous, the substance gains weight and air is consumed in the process. Lavoisier was able to repeat this experiment with a variety of substances. He sent secret memos to the science academy hypothesizing that a part of common air was contributing to combustion. From this time on, Lavoisier was convinced the phlogiston theory was bogus.

Lavoisier dined with Priestly in 1774 to discuss Priestly's discovery, "dephlogisticated air." Others had completed the same experiment and observed similar results.

Lavoisier had been observing combustion and what happened to air. He now conjectured that a part of air may be the fuel of combustion. He determined that gas released in combustion was equal to the air consumed. In other words, a component of air may be the substance that combusts. Air must be made up of more than just a single element; it must be made up of at least two parts. This was the discovery of a new substance that was also an element. In 1777 Lavoisier named it oxygen.

Henry Cavendish (British 1731-1810) an interesting son of a wealthy family lived an eccentric life of science curiosity and experimentation. An avowed phlogiston, Cavendish combined metals of iron, tin or zinc with some acids, and a very very lightweight gas was discharged. Thinking it might be "pure phlogiston," he called it "factious air." In 1777, it is later classified as hydrogen.

### The water controversy (1774-1785)

The two elements of water had been discovered, oxygen and hydrogen. With two new elements isolated, hydrogen and oxygen, they could combine to form a compound, water? This wasn't easy to accept. Water had been considered a single element for over 1,500 years. The composition of water had yet to be proven.

Earlier, an experiment had been completed combining common air with "inflammable air" (hydrogen). In 1774, John Wartire combined the gases in a bell and exploded them with an electrical spark. There were small drops of water on the side of the vessel. Others repeated the experiment–always producing small droplets of water.

A few years later, in 1781, Cavendish expanded the study by combining "dephlogisticated air" (oxygen) with "inflammable air" (hydrogen). This produced prodigious volumes of water. But when repeating the experiment with normal air, he noted that only about 1/5th of the oxygen component was used. He noted the other 4/5th did not bind into water (it was actually nitrogen). Cavendish had quietly discovered a chemical formula. With pure oxygen and hydrogen, there is a two to one ratio, two parts of hydrogen and one part of oxygen produced pure water.

Lavoisier was actually the person to explain what happened here. Clarification was his strength. Two elements combined to form water, two parts hydrogen to one part oxygen. This was a compound that was formed via a chemical reaction and can be written as a chemical formula.

In 1783, Lavoisier and his partner Laplace request a demonstration before the French academy. Feeding hydrogen and oxygen into a glass sphere, they produce water whose weight equals the weight of the two gases. Cavendish may have been the first to discover this, but Lavoisier was the one to make the convincing presentation in front of the right audience.

If you can make water, you must also be able to split the elements apart. Lavoisier demonstrated this by mixing water with iron metal. The iron oxidizes and expels pure hydrogen. He repeated this experiment with a variety of different metals, including a hot gun barrel.

The final confirmation of the makeup of the water molecule came from electro-chemistry, causing a reaction by use of an electric current. In 1800, the first battery stack was used to transmit electricity in water. When brass terminals were dipped into the water, the negative wire expelled pure hydrogen and when the positive terminal was fitted with platinum wires, it expelled pure oxygen always in perfect proportion, two parts hydrogen for one part oxygen. Electrolysis had achieved splitting water into hydrogen and oxygen with electricity.

### The Foundation of Modern Chemistry Era

Credit goes to Antoine Lavoisier for clarifying and promoting the understanding of chemistry. The modern era of chemistry was born. First he reformed the language and documented the elements known in 1789. The elements were named for what they were and the phlogiston lexicon was obliterated. The oxygen theory was completely endorsed. His book, *The Treatise of Elementary Chemistry,* was written in 1789. It conclusively defined an element, a chemical reaction and the understanding of the transformation of matter through chemical reactions. These principles of chemistry were understandable, definitive and accepted.

The discovery of the elements deserves note. Oxygen is the most prevalent element on earth, hydrogen is the 10th most common element on earth, but is the most pervasive in the Cosmos. Oxygen and hydrogen make up the water compound. Life on earth would not exist without water.

Such monumental discoveries, and yet the scientists were under recognized. The American and French Revolutions were in full swing. Joseph Priestly was loyal to both causes and consequently fell into political disfavor. His house was burned. He moved to the United States and performed research with Benjamin Franklin, only to fall ill and die in 1804.

Lavoisier was a former tax collector before his chemistry career. A disgruntled chemist indicated Lavoisier as an enemy of the people and he was guillotined in 1794.

### Lavoisier...

- Proves water is not an element (1770)
- Shows the combusted metal gains weight (1772)
- Shows that the substance in air that fuels combustion is actually an element called oxygen (1778)
- Introduces the antiphlogiston theory (1783)
- Demonstrates the composition of water, and then splits the two elements back to pure hydrogen and oxygen (1783)
- Devises new chemical nomenclature, replaced alchemist terminology (1787)
- Demonstrates the conserva-tion of matter–matter cannot be destroyed, only trans-formed in a reaction (1789)
- Updates the known elements (1789).
- Writes the treatise of modern chemistry which popularizes contemporary chemistry (1789)
- Analyzes the combustion of oxygen in organic substances
- Writes early chemical formulas

# Periodic Table

## of the Elements

| I | 2 | 3 | 4 | 5 | 6 | 7 | 8 | 9 | 10 | 11 | 12 | 13 | 14 | 15 | 16 | 17 | 18 |
|---|---|---|---|---|---|---|---|---|---|---|---|---|---|---|---|---|---|
| 1 H 1.0079 | | | | | | | | | | | | | | | | | 2 He 4.0026 |
| 3 Li 6.941 | 4 Be 9.01218 | | | | | | | | | | | 5 B 10.811 | 6 C 12.011 | 7 N 14.0067 | 8 O 15.9994 | 9 F 18.998403 | 10 Ne 20.1797 |
| 11 Na 22.98977 | 12 Mg 24.305 | | | | | | | | | | | 13 Al 26.98154 | 14 Si 28.0855 | 15 P 30.97376 | 16 S 32.066 | 17 Cl 35.4527 | 18 Ar 39.948 |
| 19 K 39.0983 | 20 Ca 40.078 | 21 Sc 44.9559 | 22 Ti 47.88 | 23 V 50.9415 | 24 Cr 51.9961 | 25 Mn 54.9380 | 26 Fe 55.847 | 27 Co 58.9332 | 28 Ni 58.69 | 29 Cu 63.546 | 30 Zn 65.39 | 31 Ga 69.723 | 32 Ge 72.61 | 33 As 74.9216 | 34 Se 78.96 | 35 Br 79.904 | 36 Kr 83.80 |
| 37 Rb 85.4678 | 38 Sr 87.62 | 39 Y 88.9059 | 40 Zr 91.224 | 41 Nb 92.9064 | 42 Mo 95.94 | 43 Tc (98.9063) | 44 Ru 101.07 | 45 Rh 102.9055 | 46 Pd 106.42 | 47 Ag 107.8682 | 48 Cd 112.411 | 49 In 114.82 | 50 Sn 118.710 | 51 Sb 121.75 | 52 Te 127.60 | 53 I 126.9045 | 54 Xe 131.29 |
| 55 Cs 132.9054 | 56 Ba 137.33 | 57 *La 138.9055 | 72 Hf 178.49 | 73 Ta 180.9479 | 74 W 183.85 | 75 Re 186.207 | 76 Os 190.2 | 77 Ir 192.22 | 78 Pt 195.08 | 79 Au 196.9665 | 80 Hg 200.59 | 81 Ti 204.3833 | 82 Pb 207.2 | 83 Bi 208.9804 | 84 Po (208.982) | 85 At (209.987) | 86 Rn (222.018) |
| 87 Fr (223.0197) | 88 Ra (226.0254) | 89 †Ac (227.0278) | 104 Rf (261.1087) | 105 Db (262.1138) | 106 Sg (263.1182) | 107 Bh (262.1229) | 108 Hs (265) | 109 Mt (266) | 110 Uun (269) | 111 Uuu (?) | 112 Uub (?) | | | | | | |

*LANTHANUM SERIES

| 58 Ce 140.115 | 59 Pr 140.9077 | 60 Nd 144.24 | 61 Pm (144.915) | 62 Sm 150.36 | 63 Eu 151.965 | 64 Gd 157.25 | 65 Tb 158.9254 | 66 Dy 162.50 | 67 Ho 164.9303 | 68 Er 167.26 | 69 Tm 168.9342 | 70 Yb 173.04 | 71 Lu 174.967 |
|---|---|---|---|---|---|---|---|---|---|---|---|---|---|

†ACTINIUM SERIES

| 90 Th (232.0381) | 91 Pr (231.0359) | 92 U 238.0289 | 93 Np (237.0482) | 94 Pu (244.064) | 95 Am (243.061) | 96 Cm (247.070) | 97 Bk (247.070) | 98 Cf (251.08) | 99 Es (252.083) | 100 Fm (257.095) | 101 Md (258.10) | 102 No (259.101) | 103 Lr (260.1053) |
|---|---|---|---|---|---|---|---|---|---|---|---|---|---|

Atomic weights conform to the 1987 values IUPAC.

# Metric and Standard Measurements

## Metric Conversions

| When You Know | Multiply By | To Find | When You Know | Multiply By | To Find |
|---|---|---|---|---|---|
| ***Length*** | | | ***Volume*** | | |
| Inches (in.) | 2.5 | Centimeters (cm) | Teaspoons (tsp.) | 5.0 | Milliliters (ml) |
| Feet (ft.) | 30.0 | Centimeters (cm) | Tablespoons (tbs.) | 15.0 | Milliliters (ml) |
| Yards (yd.) | 0.9 | Meters (m) | Fluid ounces (fl.oz.) | 30.0 | Milliliters (ml) |
| Miles (mi.) | 1.6 | Kilometers (km) | Cups (c) | 0.24 | Liters (l) |
| Centimeters (cm) | 0.4 | Inches (in.) | Pints (pt.) | 0.47 | Liters (l) |
| Meters (m) | 3.3 | Feet (ft.) | Quarts (qts.) | 0.95 | Liters (l) |
| Meters (m) | 1.09 | Yard (yd.) | Gallons (gal.) | 3.8 | Liters (l) |
| Kilometers (km) | 0.6 | Mile (mi.) | Cubic feet (ft.$^3$) | 0.03 | Cubic meters (m$^3$) |
| | | | Cubic yards (yd.$^3$) | 0.76 | Cubic meters (m$^3$) |
| ***Area*** | | | Milliliters (ml) | 0.2 | Teaspoons (tsp.) |
| Square inches (in.$^2$) | 6.5 | Square centimeters (cm$^2$) | Milliliters (ml) | 0.7 | Tablespoons (tbs.) |
| Square feet (ft.$^2$) | 0.09 | Square meters (m$^2$) | Milliliters (ml) | 0.3 | Fluid ounces (fl.oz.) |
| Square yards (yd.$^2$) | 0.84 | Square meters (m$^2$) | Liters (l) | 4.2 | Cups (c) |
| Square miles (mi.$^2$) | 2.6 | Square kilometers (km$^2$) | Liters (l) | 2.1 | Pints (pt.) |
| Acre (a.) | 0.4 | Hectares (ha) | Liters (l) | 1.06 | Quarts (qts.) |
| Square centimeter (cm$^2$) | 0.16 | Square inches (in.$^2$) | Liters (l) | 0.26 | Gallons (gal.) |
| Square meter (m$^2$) | 10.8 | Square feet (ft.$^2$) | Cubic meters (m$^3$) | 35.0 | Cubic feet (ft.$^3$) |
| Square meter (m$^2$) | 1.2 | Square yards (yd.$^2$) | Cubic meters (m$^3$) | 1.3 | Cubic yards (yd.$^3$) |
| Square kilometer (km$^2$) | 0.4 | Square miles (mi.$^2$) | | | |
| Hectare (ha) | 2.5 | Acres (a.) | ***Temperature*** | | |
| | | | Degrees Celsius (°C) | (9/5 x °C) +32 | Degrees Fahrenheit (°F) |
| ***Mass*** | | | Degrees Fahrenheit (°F) | 5/9 x (°F-32) | Degrees Celsius (°C) |
| Ounces (oz.) | 28.35 | Grams (g) | | | |
| Pound (lb.) | 0.45 | Kilograms (kg) | | | |
| Short ton (2,000 lbs.) | 0.9 | Tones-metric ton (t.) | | | |
| Grams (g) | 0.035 | Ounces (oz.) | | | |
| Kilograms (kg) | 2.2 | Pounds (lbs.) | | | |
| Tones (t.) | 1.1 | Short tons (2,000 lbs.) | | | |

# Metric and Standard Measurements

## Metric Conversions

### Flow Rate

| | |
|---|---|
| 1 gallon per minute | = $2.23 \times 10^{-3}$ cubic feet/sec. (cfs) |
| | = $4.42 \times 10^{-3}$ acre feet/day |
| | = $6.31 \times 10^{-5}$ $m^3$/day |
| | = 5.42 $m^3$/day |
| 1 cubic foot per second | = 449 gallon/min. (gpm) |
| | = 0.0283 $m^3$/sec. |
| | = 2450 $m^3$/day |
| 1 cubic meter per second | = $1.58 \times 10^{4}$ gpm |
| | = 35.3 cfs |
| | = $8.64 \times 10^{-4}$ $m^3$/sec. |
| 1 cubic meter per day | = 0.183 gpm |
| | = $4.09 \times 10^{-4}$ cfs |
| | = $1.16 \times 10^{-5}$ $m^3$/sec. |

### *Velocity*

| | |
|---|---|
| 1 foot/second | = 0.682 miles/hour |
| | = 0.3048 meters/second |
| 1 mile/hour | = 1.467 feet/second |
| | = 1.609 kilometers/hour |
| 1 meter/second | = 3.6 kilometers/hour |
| | = 3.28 feet/second |
| | = 2.237 miles/hour |
| 1 kilometer/hour | = 0.621 miles/hour |

### *Length*

| *Unit* | *Number of Meters* |
|---|---|
| Kilometer | 1,000 |
| Hectometer | 100 |
| Decameter | 10 |
| Meter | 1 |
| Decimeter | 0.1 |
| Centimeter | 0.01 |
| Millimeter | 0.001 |

### *Area*

| **Unit** | **Number of Square Meters** |
|---|---|
| Sq. kilometer | 1,000,000 |
| Hectare | 10,000 |
| Are | 100 |
| Centare | 1 |
| Sq. centimeter | 0.0001 |

### *Volume*

| **Unit** | **Number of Liters** |
|---|---|
| Kiloliters | 1,000 |
| Hectoliters | 100 |
| Decaliter | 10 |
| Liter | 1 |
| Deciliter | 0.1 |
| Centiliter | 0.01 |
| Milliliter | 0.001 |

### *Mass*

| *Unit* | ***Number of* Grams** |
|---|---|
| Metric ton or tonne | 1,000,000 |
| Kilogram | 1,000 |
| Hectogram | 100 |
| Decagram | 10 |
| Gram | 1 |
| Decigram | 0.1 |
| Centigram | 0.01 |
| Milligram | 0.001 |

# International Metric System

## Conversions

| ***Linear*** | |
|---|---|
| 10 millimeter (mm) | = 1 centimeter (cm) |
| 10 cm | = 1 decimeter (dm) |
| 10 dm | = 1 meter (m) |
| 1,000 m | = 1 kilometer |

| ***Weight*** | |
|---|---|
| 10 decigrams | = 1 gram |
| 1,000 grams | = 1 kilogram |
| 1,000 (kg) | = 1 metric ton |

| ***Area*** | |
|---|---|
| 10,000 square meters | = 1 hectare |
| 100 hectares | = 1 square kilometer |

| ***Liquid*** | |
|---|---|
| 1,000 milliliters (ml) | = 1 liter |

| ***Water Quality Unit Conversions*** | |
|---|---|
| 1 kilogram (kg) | = 1000 grams (g) |
| 1 gram (g) | = 1000 milligrams (mg) |
| 1 kilogram (kg) | = 1 million milligrams(mg) |
| 1 kilogram (kg) | = 1 billion micrograms (ìg) |
| 1 milligram (mg) | = 1000 micrograms (ìg) |

***Concentrations:***

| | |
|---|---|
| 1 gram/L | = 1 part per thousand (ppt) |
| 1 milligram/L | = 1 part per million (ppm) |
| 1 microgram/L | = 1 part per billion (ppb) |

# Glossary

**absorption.** (1) The entrance of water into the soil or rocks by all natural processes, including the infiltration of precipitation or snowmelt, gravity flow of streams into the valley alluvium, into sinkholes or other large openings, and the movement of atmospheric moisture. (2) The uptake of water or dissolved chemicals by a cell or an organism (as tree roots absorb dissolved nutrients in soil). (3) More generally, the process by which substances in gaseous, liquid, or solid form dissolve or mix with other substances.

**Acid Mine Drainage (AMD).** Drainage of water from areas that have been mined for coal or other mineral ores; the water has low pH, sometimes less than 2.0, because of its contact with sulfur-bearing material.

**acid precipitation.** Occurs where sulfur dioxide and nitrogen oxides react with oxygen, and then moisture, in the atmosphere to produce sulfuric and nitric acids.

**acidic.** A solution that has more hydrogen ions than it does hydroxide ions.

**adsorption.** The adherence of a gas, liquid, or dissolved material on the surface of a solid.

**aerobic.** Pertaining to organisms which live or are active only in the presence of oxygen. Contrast with anaerobic.

**agitation.** To stir or shake vigorously to suspend or mix particles in a solution.

**alkalinity.** A property of water that helps prevent (buffer) drastic changes in pH, thus protecting humans, wildlife, and aquatic life from the harmful effects of acid precipitation and acidic effluents.

**amino acids.** Molecules containing a protonated amino group ($NH_3^+$) and an ionized carboxyl group ($COO^-$).

**anaerobic.** Characterizing organisms able to live and grow only where there is no air or free oxygen, and conditions that exist only in the absence of air or free oxygen. Contrast with aerobic.

**anion.** A negatively charged ion.

**anthropogenic.** Involving the impact of man on nature; induced, caused, or altered by the presence of activities of man, as in water and air pollution.

**antiscalants.** Materials or liquids that resist scale formation completely.

**baseline (data).** A quantitative level or value from which other data and observations of a comparable nature are referenced. Information accumulated concerning the state of a system, process, or activity before the initiation of actions that may result in changes.

**basic.** Containing more hydroxide ions than hydrogen ions.

**benthic.** Referring to organisms that live on the bottom of water bodies.

**Best Management Practices (BMP).** Methods determined by land and water managers to describe land use measures designed to reduce or eliminate nonpoint source pollution.

**Biochemical Oxygen Demand (BOD).** The amount of oxygen consumed. Measures the oxygen removed from water during chemical reactions, such as the oxidation of sulfides, ferrous iron, and ammonia, or from biological activity.

**biodegradable.** An organic substance that is quickly broken down by normal environmental processes.

**biomass.** The weight of a living organism or assemblage of organisms.

**biopolymers.** A macromolecule in

a living organism that is formed by linking together several smaller molecules, such as protein from amino acids or DNA from nucleotides. biovailable. Able to be assimi-lated (absorbed) by organisms.

**buffer.** A solution which is resistant to pH changes, or a solution or liquid whose chemical makeup tends to neutralize acids or bases without a great change in pH. Surface waters and soils with chemical buffers are not as susceptible to acid deposition as those with poor buffering capacity. (See alkalinity).

**buffer strip.** (1) Strips of grass or other erosion-resisting vegetation between or below cultivated strips or fields. (2) Grassed or planted zones which act as a protective barrier between an area which experiences land uses that cause erosion and a water body. Also referred to as filter strips, vegetated filter strips, and grassed buffers.

**by-product.** Materials, other than the intended product, generated as a result of an industrial process.

**carbonate.** The collective term for the natural inorganic chemical compounds related to carbon dioxide that exist in natural waterways, for example, $CaCO_3$, or calcium carbonate.

**carbonic acid.** A weak, unstable acid, $H_2CO_3$, present in solutions of carbon dioxide and water. The carbonic content of natural, unpolluted rainfall lowers its pH to about 5.6.

**cation.** A positively charged ion.

**chlorophyll.** The green pigments of plants. There are seven known types of Chlorophyll; Chlorophyll a and Chlorophyll b are the two most common forms. A green photosynthetic coloring matter of plants found in chloroplasts and made up chiefly of a blue-black ester.

**coagulation.** In water treatment, the use of chemicals to make suspended solids gather or group together into small flocs.

**cold-blooded.** Organisms unable to internally control body tempera-ture but instead take on the temperature of the water in which they live.

**colloidal matter.** Any substance with particles in such a fine state of subdivision dispersed in a medium (for example, water) that they do not settle out, but not in so fine a state of subdivision that they can be said to truly dissolved.

**compounds.** (1) A substance composed of separate elements, ingredients, or parts. Water is a compound consisting of hydrogen and oxygen, chemical symbol $H_2O$. (2) A substance composed of two or more elements whose composition is constant. For example, common table salt (sodium chloride–NaCl) is a compound consisting of one atom each of sodium (Na) and Chlorine (Cl).

**conductivity.** Also know as specific conductance, is a measure of how well a water sample conducts electricity. The unit of measurement for conductivity is microsiemens/cm or micromhos/cm.

**conduits.** (1) A natural or artificial channel through which liquids may be conveyed. (2) (dam) A closed channel for conveying discharge through, under, or around a dam.

**constituents.** Any of the chemical substances found in water. Typically, measurements of such constituents in sampled drinking water may consist of Total Dissolved Solids (TDS), hardness (concentrations of calcium and magnesium, specifically), so-dium, potassium, sulfate, chlo-ride, nitrate, alkalinity, bicar-bonate, carbonate, fluoride, arsenic, iron, manganese, copper, zinc, barium, boron, silica, as well as other physical charac-ter-istics and properties such as water color, turbidity, pH, and Electro-Conductivity (EC).

**contour farming.** Involves planting crops across a hillside rather than up and down the hill. The rows of crops mimic contour lines or lines of equal elevation along the hill side. Leaving crop stubble or remnants during harvest will reduce erosion.

**corrosive.** A substance that dete-rio-rates material, such as pipe, through electrochemical processes.

**covalent bond.** A force that holds two atoms tightly to each other; found when the two atoms share one or more electron pairs.

**deionized water.** Water that has been passed through resins that remove all ions.

**denitrification.** A process which reduces nitrates to intermediate nitrites and eventually nitrogen or nitrous oxide.

**density.** (1) Matter measured as mass per unit volume; a measure of how heavy a substance (solid, liquid or gas) is for its size. The density of water is 1.0 gram per cubic centimeter or about 62.4 pounds per cubic foot. (2) (biology) The number per unit area of individuals of any given species at any given time. A term used synonymously with popula-tion density.

**detritus.** (1) The heavier mineral debris moved by natural water courses, usually in the form of bed load. (2) The sand, grit, and other coarse material removed by differential sedimentation in a relatively short period of deten-tion. (3) Bits of vegetation, animal remains, and other organic material that form the base of food chains in wetlands and many other kinds of habitats.

**dilution.** The reduction of the concentration of a substance in air or water.

**dispersants.** A chemical agent used to break up concentrations of organic material such as spilled oil on a water surface.

**dissolution.** (1) Change from a solid to a liquid state; solution by heat or moisture; liquefaction; melting. (2) Change of form by chemical agency; decomposition; resolution.

**Dissolved Oxygen (DO).** The concentration of free (not chemically combined) molecular oxygen (a gas) dissolved in water, usually expressed in milligrams per liter, parts per million, or percent of saturation. Adequate concentrations of dissolved oxygen are necessary for the life of fish and other aquatic organisms and the prevention of offensive odors. DO levels are considered the most important and commonly employed measurement of water quality and indicator of a water body's ability to support desirable aquatic life.

**diurnal.** Active during the daytime.

**ecosystem (ecology).** A community of animals, plants, and bacteria, and its interrelated physical and chemical environment. An ecosystem can be as small as a rotting log or a puddle of water, but current management efforts typically focus on larger landscape units, such as a mountain range, a river basin, or a watershed.

**effluent.** (1) Flowing out or flowing away; something that flows out or forth, especially a stream flowing out of a body of water. (2) A stream that flows out of a larger stream, a lake, or another body of water. (3) A waste liquid discharge from a manufacturing or treatment process, in its natural state or partially or completely treated, that dis-charges into the environment. The outflows from sewage or industrial plants, etc.

**estuaries.** Drainage channels adjacent to the sea, frequently the lower courses of streams, which are subject to the periodic rise and fall of tides.

**eutrophication.** Having water rich in mineral and organic nutrients that promote a proliferation of plant life, especially algae, which overproduce, die off, and the bacteria that decomposes them eventually reduces the dissolved oxygen content, sometimes causing the extinction of other organisms (fish, macroinvertebrates); typically in a lake or pond.

**fossil fuels.** (1) A fuel that contains energy locked up in chemical compounds by the plants and animals of former ages. (2) Combustible gases, liquids, and solids found in the earth's crust, resulting from the metamorpho-

sis of plants and animals living in past geologic ages. For example, coal, petroleum, and natural gas.

**granite (geology).** A light-colored plutonic igneous rock made up of interlocking grains of glassy or milky quartz, white or pink feldspar, and specks of dark mica or hornblende. The Sierra Nevada Mountains (California and Nevada) are made up of granite and similar rock types.

**hard water.** Received its name based on the fact that it is difficult (hard) to produce a soapy lather using such water. Water containing relatively high concentrations of calcium ($Ca^{2+}$), magnesium ($Mg^{2+}$), or iron (III) ($Fe^{3+}$) ions.

**heavy metal.** (1) Metallic elements with high atomic weights, e.g., mercury, chromium, cadmium, arsenic, and lead. They can damage living things at low concentrations and tend to accumulate in the food chain. (2) Those metals that have high density; in agronomic usage these include copper, iron, manganese, molybdenum, cobalt, zinc, cadmium, mercury, nickel and lead. These sub-stances are considered toxic at specified concentrations. (3) Metals having a specific gravity of 5.0 or greater; generally toxic in relatively low concentrations to plant and animal life and tend to accumulate in the food chain. Examples include lead, mercury, cadmium, chromium, and arsenic.

**hub-baffles.** Technology that uses special winged baffles on power plant turbines to pull more air into the turbines to increase DO levels of the outflow.

**humus.** (1) Organic materials resulting from decay of plant or animal matter. Organic portion of the soil remaining after prolonged microbial decomposition. (2) A brown or black organic substance consisting of partially or wholly decayed vegetable or animal matter that provides nutrients for plants and increases the ability of soil to retain water. Also referred to as compost.

**hydrocarbons.** Chemical compounds that consist entirely of carbon and hydrogen, such as petroleum, natural gas, and coal.

**hydroelectric/hydropower.** Having to do with production of electricity by water power from falling water, often from a reservoir behind a dam.

**hydrogen bond.** Attraction of one water molecule to a neighboring water molecule.

**igneous (geology).** Resulting from, or produced by, the action of fire; lavas and basalt are igneous rocks.

**impermeable.** Material that does not permit liquids to pass through.

**influent.** Water or other liquid, raw or partially flowing into a reser-voir, basin, treatment process or treatment plant.

**insoluble.** Material that cannot be dissolved: insoluble matter.

**interstices.** (1) A very small open space in a rock or granular material. Also called a void or void space. (2) The openings or pore spaces in a rock, soil, and other such material. In the zone of saturation they are filled with water. Synonymous with void or pore.

**ion exchange.** (1) The substitution of one ion for another in certain substances. Either anion exchange or cation exchange is possible. The most common cation exchange involves the conversion of hard water to soft water by means of a water softening process. Hard water contains the divalent ions of calcium ($Ca_{+}^{2}$) and magnesium ($Mg^{+2}$), which cause soap and detergents to form precipitates in water. A water softener consists of a resin that is saturated with sodium ions ($Na^{+}$). As hard water percolates through the resin, the ions of calcium or magnesium are removed as they attach to the resin, thus releasing (being exchanged for) sodium ions.

**ions.** Atoms or molecules that become charged when they either gain or lose electrons.

**Jackson Candle.** Jackson candle turbidimeter, used in 1900 by Whipple and Jackson to make the first standard measurements of turbidity. Jackson's method

involved holding a flat-bottomed, calibrated glass tube over a special candle and pouring the water sample into the tube while looking down the tube until turbidity of the sample completely blocked the view of the candle flame. The tube was calibrated with measured dilutions of a standard solution of diatomaceous earth in distilled water. This method is not sensitive enough to detect very low levels of turbidity, such as those found in treated water.

**leach.** To remove soluble or other constituents from a medium by the action of a percolating liquid, as in leaching salts from the soil by the application of water.

**legumes.** Nitrogen fixing crops.

**limestone (geology).** A sedimentary rock composed of calcite, or calcium carbonate ($CaCO_3$), and sometimes containing shells and other hard parts of prehistoric water animals and plants. When chemical conditions are right, some calcite crystallizes in sea water and settles to the bottom to form limestone.

**limiting nutrient.** Limitation in this context means that if there is too little of either of these nutrients in the water, in plant-available forms (mainly nitrate and phosphate ions), it will make further growth impossible for the algal species in question.

**limnologists.** A hydrologist who studies freshwater and the aquatic environment and its life; the study of the physical, chemical, hydrological, and biological aspects of fresh water bodies. The scientific study of conditions in freshwater lakes, ponds, and streams.

**macroinvertebrates.** Invertebrate animals (animals without backbones) large enough to be observed without the aid of a microscope or other magnification.

**Maximum Contaminant Level (MCL).** The designation given by the U.S. Environmental Protec-tion Agency (EPA) to water-quality standards supported under the Safe Drinking Water Act. The MCL is the greatest amount of a contaminant that can be present in drinking water without causing a risk to human health.

**mean (average).** The sum of a set of observations divided by the number of observations.

**metamorphic rocks (geology).** A sedimentary or igneous rock that has been changed by pressure, heat, or chemical action. For example, limestone, a sedimen-tary rock, is converted to marble, a metamorphic rock.

**methemoglobinemia.** A human blood disorder that impairs the ability of the blood supply to carry oxygen throughout the body. Also known as "blue baby syndrome", it is frequently caused by high concentrations of nitrate in drinking water supplies. It primarily affects infants less than 6 months of age. Most instances of the problem can be traced to babies drinking milk formula mixed in water with very high nitrate levels.

**microorganisms.** Very small animals and plants that are too small to be seen by the naked eye and must be observed using a microscope. Microorganisms in water include algae, bacteria, viruses, and protozoa.

**nephelometer.** A device which measures the intensity of light scattered at right angles to its path through a sample. It is used to measure turbidity, and the results are expressed in Nephelometric Turbidity Units (NTUs).

**nitrification.** (1) The biochemical transformation of ammonium nitrogen to nitrate nitrogen. (2) The conversion of nitrogenous matter into nitrates by bacteria; the process whereby ammonia in wastewater is oxidized to nitrite and then to nitrate by bacterial or chemical reactions.

**nitrogen fixation.** The conversion of element nitrogen in the atmosphere (N2) to a form (e.g., ammonia) that can be used as a nitrogen source by organ-isms. Biological nitrogen fixation

is carried out by a variety of organisms, including certain species of blue-green algae, and certain bacteria associated with legumes. In addition, the heat from lightning can fix nitrogen.

**nucleic acids.** Class of macromolecules (made up of nucleotide monomers) that contains the genetic information of organisms; DNA and RNA.

**nutrient loading.** Discharging of nutrients from the watershed (basin) into receiving water body (lake, stream, wetland); expressed usually as mass per unit area per unit time (kg/ha/yr or lbs/acre/year).

**organic matter.** Dead plant and animal matter. (See humus).

**osmotic balance.** When the movement of water in and out of the cell is equal.

**oxidize.** (1) To combine with oxygen; make into an oxide. (2) To increase the positive charge or valence of (an ele-ment) by removing electrons.

**pH.** Is generally referred to as hydrogen ion concentration of hydroxide ions when it is dissolved in water. Mathemati-cally it is defined as the negative of the logarithm of the hydrogen ion concentration.

**phosphate rocks.** Rocks with a high concentration of the mineral apatite.

**photosynthesis.** The process in green plants and certain other organisms by which carbohydrates are synthesized from carbon dioxide and water using light as an energy source. Most forms of photosynthesis release oxygen as a byproduct. Chlorphyll typically acts as the catalyst in this process.

**phreatophyte.** Water-loving plants, such as willows.

**polyacrylates.** A non-biodegradable polymer that is a super absorber (used in diapers) and an antiscalant (used to prevent build-up in boilers).

**polyaspartates.** A recently discovered biodegradable polymer that is used for the same purposes as polyacrylates, but biodegradable.

**polyvalent cations.** Ions with a positive charge of two (2) or greater.

**precipitate.** (1) An insoluble, finely divided substance which is a product of a chemical reaction within a liquid. (2) The separation from solution of an insoluble substance. (3) A solid which forms from a liquid suspension as a result of a chemical reaction. The material (floc) is insoluble in water and will settle out over time.

**remediation (environmental).** Cleanup or other methods used to remove or contain a toxic spill or hazardous materials from a Superfund site or other impacted area or watershed.

**reservoir (water).** (1) A pond, lake, or basin, either natural or artificial, for the storage, regulation, and control of water. (2) An artificially created lake in which water is collected and stored for future use. (3) Any natural or artificial storage place from which water may be withdrawn as needed.

**residence time.** The time (years) during which nutrients or sea salt will remain in a marine system as a result of internal sinks for these substances.

**resin (water treatment).** Refers to ion exchange resin products which are manufactured organic polymer beads used in softening and other ion exchange processes to remove dissolved salts from water.

**respiration (biology).** (1) The process in which an organism uses oxygen for its life processes and gives off carbon dioxide. (2) The oxidative process occurring within living cells by which the chemi-cal energy of organic molecules (i.e., substances containing carbon, hydrogen, and oxygen) is released in a series of metabolic steps involving the consumption of oxygen ($O_2$) and the liberation of carbon dioxide ($CO_2$) and water ($H_2O$)

**reverse osmosis.** (1) A filtration process which forces water through membranes that contain holes so small that even salts cannot pass through. Reverse osmosis removes microorganisms, organic chemicals, and inorganic chemicals, producing very pure water. Reverse osmosis units require regular maintenance or they can become a health hazard. (2) (desalination) Refers to the process of removing salts from water using a membrane. With reverse osmosis, the product water passes through a fine membrane that the salts are unable to pass through, while the salt waste (brine) is removed and disposed. This process differs from electrodialysis, where the salts are extracted from the feedwater by using a membrane with an electrical current to separate the ions. The positive ions go through one membrane, while the negative ions flow through a different membrane, leaving the end product of freshwater. (3) (water quality) An advanced method of water or wastewater treatment that relies on a semi-permeable membrane to separate waters from pollutants. An external force is used to reverse the normal osmotic process resulting in the solvent moving from a solution of higher concentration to one of lower concentration.

**riparian.** Pertaining to the banks of a river, stream, waterway, or other, typically, flowing body of water as well as to plant and animal communities along such bodies of water. This term is also commonly used for other bodies of water, e.g., ponds, lakes, etc., although littoral is the more precise term for such stationary bodies of water. Also refers to the legal doctrine (Riparian Doctrine and Riparian Water Rights) that says a property owner along the banks of a surface water body has the primary right to withdraw water for reasonable use.

**runoff.** That portion of precipitation which is not intercepted by vegetation, absorbed by the land surface or evaporated, and thus flows overland into a depression, stream, lake or ocean. Often carries sediments or pollutants.

**salinity.** (1) The concentration of dissolved salts in water or soil water. Salinity may be expressed in terms of a concentration or as an electrical conductivity. When describing salinity influenced by seawater, salinity often refers to the concentration of chlorides in the water. (2) The relative concentration of salts, usually sodium chloride, in a given water sample. It is usually expressed in terms of the num-ber of parts per thousand (ppt) or parts per million (ppm) of chloride (Cl). Although the measurement takes into account all of the dissolved salts, sodium chloride (NaCl) normally constitutes the primary salt being measured. Salinity can harm many plants, causing leaves to scorch and turn yellow and stunting plant growth. As a reference, the salinity of seawa-ter is approximately 35 ppt or 35,000 ppm.

**saturated.** (1) Generally, filled to capacity; having absorbed all that can be taken up; soaked through with moisture. (2) (hydrologic) A condition often used in reference to soils in which all voids or pore spaces between soil particles are filled with water. (3) (chemistry) Describes a solution in its most concentrated state in which dissolved material can remain in solution under given conditions of temperature, pressure, etc.

**scale.** (1) An accumulation of solid material on interior surfaces, such as pipelines, tanks, and boilers, as a result of the precipi-tation of mineral salts from water, most typically salts of calcium. Hard water leaves a deposit (scale) in steam irons, coffee makers, and water heaters. (2) A hard incrustation usually rich in sulfate of calcium that is deposited on the inside of a vessel (as a hot water heater) in which water is heated.

**Secondary Drinking Water Standard.** Non-enforceable standards related to the aesthetic quality of drinking water such as those relating to taste and odor; generally set by the U.S. Environ-mental Protection Agency

(EPA) or state water-quality enforce-ment agencies based on EPA guidance.

**sediment.** Created from natural process of erosion, where wind, water, frost and ice slowly break down rocks into finer and finer pieces. Runoff often carries sediment into nearby waterways.

**sedimentary rock (geology).** (1) Rocks formed by the accumulation of sediment in water (aqueous deposits) or from air (eolian deposits). A characteristic feature of sedimentary deposits is a layered structure known as stratification or bedding. (2) Many sedimentary rocks show distinct layering, which is the result of different types of sediment being deposited in succession.

**sedimentation flux.** Change in the amount of sediment transported.

**seepage.** (1) The percolation of water through the soil from unlined channels, ditches, watercourses and water storage facilities. (2) The passage of water or other fluid through a porous medium, such as the passage of water through an earth embankment or masonry wall. (3) Ground water emerging on the face of a stream bank. Seepage is generally expressed as flow volume per unit of time.

**silica (geology).** Silicon dioxide ($SiO_2$). It occurs in crystalline (quartz), amorphous (opal), or impure (silica sand) forms.

**sludge.** (1) The concentrated solid material which has been col-lected in treatment plants. Sludge must be treated and disposed of with minimum pollution of air, land or water. Also, the settleable solids sepa-rated from water during process-ing. (2) Semisolid material such as the type precipitated by a sewage treatment plant. The terms biosolids, sludge, and sewage sludge can be used interchangeably. (3) (water quality) Mud, mire, or ooze covering the ground or forming a deposit, as on a riverbed.

**solar radiation.** Radiation from the sun. The energy force that drives the hydrologic cycle.

**solubility.** The relative capacity of a substance to serve as a solute. Sugar has high solubility in water, whereas gold has low solubility in water.

**species shift.** A shift or change in species, usually due to environmental stressors.

**specific heat.** The amount of heat per unit mass required to raise the temperature of a material by one degree Celsius ($^0C$).

**stewardship (ecology).** Caring for land and associated resources and maintaining healthy ecosystems for future generations.

**stratification.** (1) The formation of separate layers (of temperature, plant, or animal life) in a lake or reservoir. Each layer has similar characteristics such as all water in the layer has the same temperature. (2) The arrangement of a body of water, such as a lake, into two or more horizontal layers of differing characteristics, such as temperature, density, etc. Also applies to other substances such as soil and snow, etc.

**surface area (lake or impoundment).** The extent of a 2-dimensional surface enclosed within a boundary.

**surfactants (water quality).** (1) The active agent in detergents that possesses a high cleaning ability. (2) An agent that is used to decrease the surface tension of water, useful for removing or dispersing oils or oily residues. Most detergents are surfactants. The term is derived from "surface active agent".

**thermal pollution.** The influx of heated water, usually from a power plant, wastewater from a factory or sewage treatment plant, or the discharge of other industrial cooling water, into a stream, lake, bay, or ocean, disturbing the tempera-ture of the given body of water. The resulting shift to a warmer aquatic environment can cause a change in species composition

and lower the dissolved oxygen content of the water. Also has application to air, through waste heat emitted by industry, home appliances, machines, etc.

**titration (chemistry).** (1) A method, or the process, of determining the strength of a solution, or the concentration of a substance in solution, in terms of the smallest amount of it required to bring about a given effect in reaction with another known solution or substance, as in the neutralization of an acid by a base. (2) A process whereby a solution of known strength (the titrant) is added to a certain volume of treated sample containing an indicator. A color change shows when the reaction is complete (the end point).

**turbidimeter.** A device used to measure the degree of turbidity, or the density of suspended solids in a sample.

**turbidity.** The cloudy appearance of water caused by the presence of suspended and colloidal matter. Technically, turbidity is an optical property of the water based on the amount of light reflected by suspended particles.

**urban runoff.** That portion of precipitation that does not naturally percolate into the ground or evaporate, but flows via overland flow, underflow, or channels or is piped into a defined surface water channel or a constructed infiltration facility from urban areas. This water is often of higher temperature and often contains oils, fertilizers, sediments, and other contaminants.

**voltmeter.** A voltmeter, also known as a voltage meter, is an instru-ment used for measuring poten-tial difference, or voltage, between two points in an electrical or electronic circuit.

**voltage.** (1) The rate at which energy is drawn from a source that produces a flow of electricity in a circuit; expressed in volts. (2) The difference in electrical potential between two points in a circuit expressed in volts.

**weathering (geology).** (1) The physical disintegration or chemical decomposition of rock due to wind, rain, heat, freezing, thawing, etc. (2) The breakdown of rock through a combination of chemical, physical, geological, and biological processes. The ultimate outcome is the generation of soil.

**wetlands.** (1) An area that is periodically inundated or satu-rated by surface or ground water on an annual or seasonal basis, that displays hydric soils, and that typically supports or is capable of supporting hydro-phytic vegetation. (2) Any number of tidal and non-tidal areas characterized by saturated or nearly saturated soils most of the year that form an interface between terrestrial (land-based) and aquatic environments; include freshwater marshes around ponds and channels (rivers and streams), brackish and salt marshes; other common names include swamps and bogs.

**zooplankton.** (1) Small, usually microscopic animals (such as protozoans) found in lakes and reservoirs that possess little or no means of propulsion. Consequently, animals belonging to this class drift along with the currents. (2) Zooplankton are capable of extensive movements within the water column and are often large enough to be seen with the unaided eye. Zooplank-ton are secondary consumers feeding upon bacteria, phy-toplankton, and detritus. Because they are the grazers in the aquatic environment, the zooplankton are a vital part of the aquatic food web. The zooplankton community is dominated by small crustaceans and rotifers.

# Index

# Milestones
## in Water Quality Management

From the beginning, humans have been challenged with finding and providing clean, healthy water to drink. Through the millennia there have been many contributions to the science and management of water quality, both for human consumption as well as in the environment. This list provides a small sample of the contributions and milestones in the history of water quality management.

**4000 B.C.** Ancient Sanskrit and Greek writings recommend water treatments like filtering through charcoal, exposing to sunlight, boiling, and straining

**1500 B.C.** Egyptians used alum to settle out suspended particles in water

**400 B.C.** Hippocrates states importance of water quality to health, recommends boiling and straining water

**312 B.C.** Start of Roman aqueduct construction

**144 B.C.** Aqua Marcis, the longest Roman aqueduct built

**1652 A.D.** First incorporated waterworks formed in Boston

**1700s** Filtration found effective for removing particles suspended in water

**1774** Chlorine is discovered in Sweden

**1800s** Slow sand filtration used in Europe

**1804** The first municipal water filtration works opens in Paisley, Scotland

**1835** Chlorine is first applied to drinking water to control foul odors in the water

**1849** Cholera epidemic–8,000 lives claimed in New York City and 5,000 in New Orleans

**1850** Swamp Act–Encourages draining of wetlands for development

**1854** Dr. John Snow discovers cholera outbreak in London is due to a contaminated well on Broad Street

**1862** Homestead Act–Opens the West to settlement and water development

**1877** Louis Pasteur develops the theory that germs spread disease

**1880s** Louis Pasteur demonstrated the "germ theory" of disease–how microbes transmit disease through water

**1882** Filtration of London drinking water begins

**1890s** Chlorine is proven an effective disinfectant of drinking water

**1890** National Weather Service–Collects data to monitor weather patterns, monitors weather data

**1894** Carey Irrigation Act–Grants public lands to states for irrigation

**1896** Louisville Water Company creates new treatment technique by combining coagula-tion with rapid-sand filtration

**1899** Rivers and Harbors Act–Prohibits discharge

| | |
|---|---|
| | of solids into navigable rivers |
| **1900s** | U.S. drinking water treatment systems built to reduce turbidity and the microbial contaminants associated with sediment; use slow sand filtration |
| **1902** | Belgium implements the first continuous use of chlorine to make drinking water biologically "safe" |
| **1906** | General Dam Act–Regulates private dam construction on navigable streams |
| **1908** | U.S. public water supply is chlorinated for the first time at Boonton reservoir supply, Jersey City, NJ |
| **1912** | Congress passes the Public Health Service Act, authorizes surveys and studies for water pollution |
| **1914** | First standards under the Public Health Service Act become law, established maxi-mum contaminant limit for drinking water |
| **1924** | Oil Pollution Act–Prohibits the discharge of oil into marine waters |
| **1925** | Rivers and Harbors Act–Authorizes the U.S. Army Corps of Engineers to survey all navigable waters and formulate general water use plans |
| **1925** | Public Health Service–Revises standards for drinking water |
| **1936** | Flood Control Act–First nationwide flood control act; introduces cost-benefit analysis |
| **1944** | Flood Control Act–Recognizes the priority of flood control over irrigation, recreation, and power production |
| **1946** | Public Health Service–Again revises standards for drinking water |
| **1948** | Water Pollution Control Act–Provides technical assistance to municipalities for wastewater treatment |
| **1954** | Small Watershed Act–Sets up a small watershed program under the Soil Conservation Service (now Natural Resources Conservation Service) |
| **1955** | An infectious hepatitis epidemic in New Delhi, India is traced to inadequately chlorinated water at a treatment plant; 1 million people infected |
| **1956** | Federal Water Pollution Control Act–Increases federal assistance for wastewater treatment |
| **1958** | Fish and Wildlife Coordination Act–Requires equal consideration of wildlife protection at federal water projects |
| **1962** | U.S. Public Health Service Drinking Water Standards Revision is accepted as minimum standards for all public water suppliers |
| **1962** | Water Resource Research Act–Establishes water resource research institutes in each state and territory with a $100,000 grant |
| **1965** | Water Resource Planning Act–Establishes the Federal Water Pollution Control Commission |
| **1960s (late)** | Industrial and agricultural advances and creation of man-made chemicals have negative impacts on environment and public health |
| **1966** | Endangered Species Act–Protects endangered species |
| **1969** | U.S. Public Health Service Community Water Supply study reveals major deficiencies in the nation's public water supplies |
| **1969** | Izaak Walton League–Establishes SOS (Save Our Streams) program; river and stream monitoring |
| **1971-78** | Maine, Minnesota, Michigan, and New Hampshire–Initiate statewide lake monitoring programs |

**1972** Federal Water Pollution Control Act Amendments–Institutes a national permit system for point source discharges; puts U.S. Army Corps of Engineers in charge of regulating discharge of dredge and fill materials

**1973** U.S. Congress debates new legislative proposals for federal safe drinking water law due to studies on the Mississippi River

**1974** Safe Drinking Water Act–Coordinates monitoring and training for safe drinking water; sets up drinking-water standards

**1977** Safe Drinking Water Act is amended to extend authorization for technical assistance information, training, and grants to the states

**1977** Clean Water Act amendment–Authorizes more grant money for states

**1980's** Improvements made in membrane development for reverse osmosis filtration and ozonation

**1981** Clean Water Act amendment–Authorizes more grant money for states

**1985** Food Security Act–Establishes erosion control programs for agricultural lands; denies federal farm benefits to farmers harvesting from converted wetlands

**1985** Rhode Island and states surrounding the Chesapeake Bay–Initiate estuary monitoring

**1986** Safe Drinking Water Act–Further amended to set mandatory deadlines for the regulation of key contaminants

**1987** Wild and Scenic Rivers Act–Protects instream flows for rivers designated wild and scenic

**1987** Clean Water Act amendment–Authorizes more grant money for states

**1987** Water Quality Act–Requires EPA to regulate storm water runoff and states to prepare nonpoint source management programs

**1988** First National Volunteer Monitoring Conference–85 participants

**1989** First Issue of The Volunteer Monitor–8 pages, 3000 copies distributed

**1994** Volunteer Water Monitors–517 Programs in 45 states; The Volunteer Monitor: 24 pages, 20,000 copies distributed

**1996** Safe Drinking Water Act–Reauthorized

**1998** Volunteer Water Monitors–800+ programs in all 50 states; The Volunteer Monitor: 24 pages, 40,000 copies distributed

**2000** Centers for Disease Control and Prevention and the National Academy of Engineering–Named water treatment as one of the most significant public health advancements of the 20th Century

**2001** More than 90% of the U.S. population is served by community water systems

**References:**


National Drinking Water Clearinghouse. 2001. *On Tap.* Retrieved February 15, 2002 from www.ndws.wvu.edu

The Watercourse. 1995. *Project WET Curriculum and Activity Guide.* Bozeman, MT: The Watercourse.

United States Environmental Protection Agency. *The History of Drinking Water Treatment.* February 2000. Retrieved February 15, 2002 from http://www.epa.gov/safewater/sdwa/trends.html

Weise, James. No date available. *Historic Milestones in Drinking Water History.* Retrieved February 16, 2002 from http://www.state.ak.us/dec/deh/water/history.htm